JANE AND JOHNNY LOVE MATH:

Recognizing and Encouraging Mathematical Talent In Elementary Students

A Guidebook for Educators and Parents

Ann E. Lupkowski
and
Susan G. Assouline

Fourth Printing

Royal Fireworks Press

Unionville, New York
Toronto, Ontario

To Mom, Dad, Peter, Debbie, and Mike, with loving thanks for your support in this and so many other projects.

A. E. L.

To José, Jason, Sonja, Mom, and Dad. Thank you for your continuous inspiration and loving support for all that I do.

S. G. A.

Copyright © 1997, R Fireworks Printing Co., Ltd.
All Rights Reserved.

Originally published in 1992 by Trillium Press, Inc.

Royal Fireworks Press
First Avenue, PO Box 399
Unionville, NY 10988
(845) 726-4444
FAX (845) 726-3824
email: rfpress@frontiernet.net

ISBN: 0-89824-539-7

Printed in the United States of America on acid-free, recycled paper using soy-based inks by the Royal Fireworks Printing Company of Unionville New York.

TABLE OF CONTENTS

Foreword iv

Introduction vi

1   OVERVIEW 1

Rationale and Philosophy 1
Historical Roots 2
Program Models: Making Informed
    Educational Decisions 3
Curriculum and Materials 6
Case Studies 7
Assessing Children's Educational Needs 7
Planning and Resources 8
Appendices, Glossary, and Index 8

2   HISTORICAL AND CURRENT PERSPECTIVES 9

Study of Mathematically Precocious Youth 11
Giftedness and Mathematical Talent 12
Conclusion 14

3   MAKING INFORMED EDUCATIONAL DECISIONS 17

Historical Background for Age-in-Grade Grouping ... 17
What is Differentiation? 18
What are the Options? 19
Comments Concerning Acceleration 27
Conclusion 31

Table of Contents

4 DIAGNOSTIC TESTING FOLLOWED BY PRESCRIPTIVE INSTRUCTION: SMPY'S DT→PI MODEL ... 33

    Overview ... 33
    Beyond-Level Assessment ... 33
    Diagnostic Testing followed by
        Prescriptive Instruction ... 34
    The Role of the Mentor ... 41
    Using the DT→PI Model ... 43
    Conclusion ... 52

5 MATHEMATICS CURRICULUM AND MATERIALS ... 55

    Overview ... 55
    What Happens in a Typical Classroom? ... 57
    NCTM Standards ... 58
    Concepts vs. Computation ... 59
    Ten Areas of Mathematics for
        Talented Elementary Students ... 60
    Materials ... 68
    Contests ... 78
    Conclusion ... 78

6 CASE STUDIES OF TALENTED YOUTHS ... 81

    Cartoon (Artist: Alejandra Piniella) ... 81
    Josh: Implementation of the DT→PI Model ... 83
    Matthew: Long Term Application
        of the DT→PI Model ... 87
    Jeff: A Bright Student Who Doesn't Need
        an Individualized Program ... 92
    Eddie: In the Midst of a Parent/School Conflict ... 93
    Elizabeth: A Mathematically Talented Girl Whose
        Parents Don't Want Her to be Different ... 101

      Steve: Radical Acceleration of a
         First-Grader *103*
      Peter: An Extraordinarily Talented Student *107*
      Conclusion *110*

7 GUIDELINES FOR A USEFUL EDUCATIONAL ASSESSMENT 111

      Cartoon (Artist: Tony Hall) *111*
      Overview *111*
      Intelligence Tests *112*
      Aptitude Tests *116*
      Achievement Tests *119*
      The Psychoeducational Report *122*

8 LONG-TERM PLANNING 129

      Conclusion *134*

Appendix A   *Forms to Aid Educators and Parents* *135*

Appendix B   *Mathematical Olympiads for*
              Elementary Schools *139*

Appendix C   *The Constructivist Approach to*
              Educating Mathematically
              Talented Students *157*

Appendix D   *Programs and Resources* *159*

Glossary *169*
References *175*
Index *185*
About the Authors *194*

Foreword

A GREATLY NEEDED HANDBOOK ABOUT YOUTHS WHO REASON WELL MATHEMATICALLY

Here, at last, parents and teachers of young math "whizzes" can find a comprehensive approach concerning identifying and educationally facilitating their precocious progeny and students. As I have learned from the Study of Mathematically Precocious Youth (SMPY) since starting it at Johns Hopkins University in 1971, an appreciable number of young boys and girls move ahead in mathematics concepts and computation far faster than the pace of their age-in-grade class. The percentage of such *wunderkind* is small, but even the top 1 in 1000 of a given year group nationwide is about 3500 youths. They constitute an invaluable natural resource that can all too easily be wasted by neglect. Drs. Assouline and Lupkowski, former postdoctoral fellows at SMPY, have addressed this situation and less selective ones squarely, in breadth and depth.

They are not concerned much with calculating prodigies, but rather with bright youngsters up to about age 11 and the sixth grade who have the potential truly to excel in mathematics and related subjects such as physics, computer science, or electrical engineering. From these will come many of the nation's top-level scientists and mathematicians. Perhaps the greatest virtue of their book is that it extends downward the vast experiences of SMPY and related groups with older youths, typically those found when they are 12 or 13 years old and in the seventh or eighth grade. The earlier a mathematically precocious student can be helped to avoid boredom and frustration caused by his or her being "out of sync" with agemates academically, the greater the likelihood that tremendous intellectual potential can be utilized effectively for the benefit of both the youth and society.

Parents and teachers who truly want to help their math-talented children or students will be well-advised to read this book carefully with pen and paper in hand to note relevant points. Then they can consult it further as a guide. Also, the many references provided should enable them to dip deeper into the literature to answer questions that arise in their minds.

Foreword

Most persons will not encounter many 1-in-1000 math reasoners. Nevertheless, the book's content is broad enough for a much wider audience, because its precepts apply to the top 5 percent or even the top 10 percent of students. Also, the authors' discussions can enrich parenting and teaching even more broadly than that. This should be an excellent main or supplemental textbook in a variety of courses training teachers for kindergarten, primary, and upper elementary schools. All teachers of the gifted will benefit much, too.

We are indebted to these able, experienced educational and school psychologists for this treatment of a long-neglected area. Of course, theirs is not the last word. Much further experience and research at these age levels will be essential. Probably, Drs. Assouline and Lupkowski can be counted on for some of that. You readers have a great opportunity to make the knowledge base even more secure.

> Julian C. Stanley, Director
> Study of Mathematically Precocious Youth
> Johns Hopkins University
> Baltimore, MD 21218

INTRODUCTION

As postdoctoral fellows with Dr. Julian C. Stanley's Study of Mathematically Precocious Youth (SMPY) at Johns Hopkins University in the late 1980's, we were surrounded by helpful information for mathematically talented youngsters. Most of the resources and programs, however, had been developed for junior high and high school students. Often, parents of children younger than junior high age called on SMPY for advice in planning their child's mathematics education. We were frustrated with telling these parents that they had to "wait a few years" before their child could participate in a formal program that was appropriate for exceptionally mathematically talented youngsters. As time went on, with the generous assistance of Dr. Stanley, we realized that his model of mathematics instruction for junior high and high school students could be successfully adapted for younger mathematically talented students. Indeed, the Center for Academic Precocity at Arizona State University and the Center for the Advancement of Academically Talented Youth at Johns Hopkins University have successfully adapted and implemented Stanley's model in their classes for elementary-age students (Daniel & Cox, 1988; Moore & Wood, 1988). Many parents and teachers who did not live within commuting distance of these programs contacted us for advice about their children. This handbook is an outgrowth of our work with those individuals and the many others we've helped in subsequent years.

SMPY's goals are Discovery, Description, Development, and Dissemination. By seeking students who are talented in mathematics (Discovery), studying them and describing their talents (Description), facilitating their education (Development), and distributing the findings (Dissemination), SMPY has influenced the academic decisions of its students (see Stanley & Benbow, 1986). For example, a large number of SMPY's students have participated in prestigious programs and competitions, won many awards, and/or graduated early from college with advanced degrees (Brody, Lupkowski, & Stanley, 1988; Stanley, 1988).

Just as there were no precedents for the type of educational decisions that needed to be made for junior high youths demonstrating extreme talent in mathematics when Dr. Stanley began SMPY in the

Introduction

early 1970's, today there are few guidelines for making educational decisions for mathematically talented elementary students. In the 1970's and 1980's, Dr. Stanley and his associates forged ahead by exploring educational alternatives and the degree to which programs or methods allowed students to proceed as rapidly as their abilities and interests would allow. The SMPY model was developed based upon that information. Our purpose is to expand upon SMPY's goals and apply the model to elementary students.

Our goal in writing this book is to present our perspective on nurturing students younger than junior high age who are talented in mathematics. We are using what has been learned in well-established programs for junior high and high school students to guide us in our treatment of and recommendations for able elementary students. For example, when SMPY was first founded by Dr. Stanley, he chose to concentrate on students in junior high--those who were preparing to study algebra. By participating in SMPY and other programs like it, talented junior high school students were able to study algebra and more advanced mathematics at an appropriate pace, which was faster than is typical. Because of SMPY and other similar programs, a wide variety of educational opportunities have become available throughout the United States to students in 7th grade and older.

We have based the ideas set forth in this book on the sound practices advocated by Dr. Stanley and his colleagues. Our extensions of this work have been tested in many ways. Two programs, the Julian C. Stanley Mentor Program (coordinated from 1989-1991 by the Study of Mathematically Precocious Youth at the University of North Texas and from 1992 to present at Carnegie Mellon University) and the Investigation of Mathematically Advanced Elementary Students (at The Connie Belin National Center for Gifted Education at The University of Iowa) serve as prototypes for schools and individuals interested in implementing mentor-paced programs for mathematically talented youth (Lupkowski, Assouline, & Vestal, 1992).

PLANS FOR THE SERIES

Jane and Johnny Love Math: Recognizing and Encouraging Mathematical Talent in Elementary Students was developed as a guidebook for parents and teachers of mathematically talented elementary school students. This book delineates our philosophy as well as methods of addressing the educational needs of mathematically

Introduction

talented students younger than age 12.

To assist parents and teachers further, we are developing a set of curricular materials that will be challenging to those talented youngsters. The first three workbooks in the *Jane and Johnny Love Math* series are in progress and nearing completion. We hope to add more workbooks and other materials to this series.

ACKNOWLEDGEMENTS

Anonymous sponsors provided the financial support for much of our work. They have also offered generous moral support for our projects on behalf of talented youth.

As we gathered material to write this book, we called upon the assistance of many people. We appreciate the valuable comments and advice given to us by Dr. Mark Saul who was especially helpful in recommending computer programs and games. We thank Dr. George Lenchner for giving us permission to reprint problems from the Mathematical Olympiads for Elementary Students. Others who assisted us in various ways in gathering materials for the book and preparing the manuscript include: Susan Anderson, Celeste Carter, Sharon Emerson, Mary Landers, Cindy Sewell, Ann Taylor, and Marjorie Whitmore. DeAnn L. Ambroson graciously and patiently proofread and formatted the entire book.

We appreciate the support provided by Dr. Nicholas Colangelo, Director of The Connie Belin National Center for Gifted Education, at The University of Iowa. We aspire to meet his high standards of excellence.

We are extremely grateful to Julian Stanley for the help and encouragement he gave us throughout the process of writing this book. He brought us together as postdoctoral fellows at the Study of Mathematically Precocious Youth at Johns Hopkins University, and his example of scholarship has set a high standard for us to follow. Julian has been both a mentor and a friend.

JANE AND JOHNNY LOVE MATH:

Recognizing and Encouraging Mathematical Talent In Elementary Students

A Guidebook for Educators and Parents

OVERVIEW

RATIONALE AND PHILOSOPHY

Eight-year-old Johnny has an IQ of 145 and seems capable of grasping anything of interest to him. Lately, he has been begging his mother to get geometry books for him. Just as Johnny's mother wonders whether or not he is ready for geometry--or even beginning algebra--his classroom teacher may be uncertain of how to design an appropriate mathematics curriculum for him. Perhaps you are the parent or teacher of a child such as Johnny and have wondered how best to help that child educationally. Our intent in writing this book is to provide some strategies for elementary students who are extremely talented in mathematics.

Systematic educational opportunities for younger children who are talented in mathematics are rare. Although a few programs for mathematically talented elementary students exist in the United States, children who do not live within commuting distance of those programs generally cannot participate. In addition to the fact that few programs exist, there are no guidelines for educators who want to develop individualized programs for young students who are extremely talented in mathematics.

The students for whom this book is written are those who show evidence of exceptional mathematical talent. Like Johnny, they might ask for mathematics books at a level far beyond what they are currently studying in school. Often they have shown an interest in numbers from an early age. For example, while pretending to lift weights, 3-year-old Chris counted the number of times the weights were lifted up to 100, showing both the ability to count to 100 and the concept of one-to-one correspondence--unusual abilities for a preschooler. As another example, seven-year-old Jane is about to complete a second book in pre-algebra, and her parents are uncertain of what to do next. Parents of children like Johnny, Chris, and Jane can list many behaviors such as these. However, simply providing examples or listing characteristics of mathematically talented children will not ensure that programs matching their needs will be developed.

GIFTEDNESS AND MATHEMATICAL TALENT

Four theoretical perspectives useful in understanding students who are talented in mathematics are: (a) the **psychometric approach**, the approach typically identified with Louis Terman and the use of IQ as a measure of giftedness; (b) the **cognitive developmental approach**, which is often identified with Piaget's stage theory of child development; (c) **information processing** which, in the field of gifted education, is identified with Sternberg's description of the dynamics of mental processes; and (d) Gardner's **multiple intelligences**. These four perspectives provide a framework from which we have built our model for educating mathematically talented elementary students. They are not mutually exclusive as theoretical perspectives and each plays an important role in helping educators understand mathematically talented youngsters. Each of these theoretical perspectives is presented in greater detail in the next chapter.

HISTORICAL ROOTS

The gifted education movement is characterized by the swinging pendulum that highlights the extremes of any movement. Following World War II and especially after Sputnik was launched in 1957, the American public and its policy makers became interested in improving education for gifted students, who were seen as a valuable national resource. However, this interest waned during the Civil Rights movement of the 1960's. The pendulum swung slowly in the opposite direction during the 1970's with the publication of the Marland Report (Marland, 1972), which stated that gifted children were underserved at the local, state, and federal level. Following that report, the Federal Office of Gifted and Talented Education was opened, and national interest in providing services for gifted students increased, although this interest was not reflected in the implementation of programs. The Office was closed in 1982; however, the passage of the Jacob K. Javits Gifted and Talented Students Education Act of 1988 reestablished a Federal Office of Gifted and Talented Education and provided funds on a competitive basis to schools and universities for developing programs and conducting research with gifted students. The gifted education movement is currently on the upswing of the pendulum; program implementation is catching up to recommendations set forth in policy statements made in previous decades. Excellent historical perspectives of the gifted education movement are provided by DeLeon and VandenBos (1985), Hildreth (1966), Tannenbaum (1979), and Whitmore (1980).

Despite the optimism associated with the progress that has been made since the 1970's, continued progress toward accurate and appropriate planning for individual children is hindered by beliefs that dictate educational policy. A major concern revolves around the issue of providing children with material that is intellectually stimulating, albeit several grade levels above materials used by the typical student of a given chronological age. One purpose of our book is to understand this concern with respect to mathematics education for extremely talented students.

PROGRAM MODELS:
MAKING INFORMED EDUCATIONAL DECISIONS

> . . . it is not enough for us to be simply awed by and appreciative of the amazing amount of precocity in mathematics and science which exists. We must ponder the question how to change our traditional educational system to accommodate these bright and able youth. Today our elementary and secondary schools do not provide the freedom necessary to exercise such exceptional individual talents. (Fox, 1974, p. 67)

Four different models are used to provide mathematics instruction for talented elementary school students: (1) the unified approach, (2) studying mathematics with older students, (3) enrichment, and (4) studying advanced content at a fast pace. Each one of these models has merits and drawbacks which are discussed below.

UNIFIED APPROACH

The unified approach, which uses the spiral curriculum, is currently the most common in elementary schools in the United States.

Overview

The defining characteristic of this approach is repetition of information throughout the elementary and middle school years. The aim is to introduce a topic and then cyclically review and further develop it. Fourth graders, for example, spend several weeks at the beginning of each school year reviewing basic computational skills. Ideally, a unified approach has not only a spiral, repetitive aspect, but also an integrative aspect. In other words, students study *real* algebra and *real* geometry throughout their years in elementary school. Europeans will tell you that they learned about algebraic functions and factors before junior high school. Americans will tell you that they finally learned functions and factors just about the time when other things (e.g., socialization with the opposite sex, etc.) became a primary reason for attending school.

One example of a unified, spiral approach that has been successfully implemented in the United States is known as Elements of Mathematics (EM). This program was designed for the top 5-10 percent of mathematics students and was developed for use in grades 7-12 (Wheatley, 1988). Because of its emphasis on secondary students, however, it is not applicable to younger students. This program requires that mathematics teachers be able to deal with very sophisticated mathematics, often a level of mathematics not presently required of mathematics education majors. This approach is heavily dependent upon the student remaining in one school district throughout grades 7-12. Considering that families often move from one city to another, this requirement becomes a liability to the program.

STUDYING MATHEMATICS WITH OLDER STUDENTS

Some schools provide for their mathematically talented students by having them simply skip ahead, either by skipping an entire grade or by taking only mathematics classes with older students. This is an attempt to provide more advanced content for students who have already mastered the content for their age level. Grade-skipping may be an appropriate option for above-average students with fairly equal strengths in math, reading, and social skills. However, for those with extreme talent in one area, such as mathematics, subject-matter acceleration may be preferred. This means moving students ahead in their area of strength, e.g., placing them in the math group in a higher grade while they remain with their age peers for all other subjects. This seems to be a reasonable alternative; however, for students extremely talented in mathematics the pace is generally too slow, even though they are placed with older students. In addition to an inappropriate pace, the material is not presented at a level that is sufficiently sophisticated for their talents. An additional concern about grade-skipping and subject matter acceleration is frequently voiced by administrators and teachers: young students might be adversely affected by being placed in a situation with older students; specifically, this approach might have negative social or emotional effects.

ENRICHMENT

Most currently available gifted programs typically prescribe enrichment activities for identified children. The intent of enrichment programs is to provide learning opportunities at a higher level and in greater depth or breadth as compared to those experiences provided in the regular classroom. However, in practice, most enrichment programs do not provide for the specific curricular needs of exceptionally able students. The typical enrichment program does not tailor activities to mathematically advanced students. For example, the gifted program in one large school district is offered only through the language arts courses. No provisions are made to enrich or accelerate the mathematics program.

In many enrichment programs, talented students are required to work at the same pace as the rest of the class and are provided with "enrichment" activities in an attempt to meet their needs for more challenging material. As a result, talented children with an aptitude in a specific area such as mathematics are forced to trudge through the curriculum at a pace more appropriate for those with average capabilities in mathematics. This denies them the opportunity for learning those advanced materials for which they are ready, and studying them at an appropriate pace.

Probably because of the emphasis the National Council of Teachers of Mathematics (NCTM) placed on it, the buzzword of the 1980's was problem solving. Problem solving was seen as an appropriate form of enrichment for gifted students. Learning problem solving skills is an important task for gifted students, and will continue to be a part of educational objectives well into the 21st century. Problem solving allows students to use their strengths in analysis and encourages divergent thinking. By itself, however, it is not sufficient as a curriculum for educating youths who are extremely talented mathematically.

Overview

The buzzword for the early 1990's is cooperative learning. Cooperative learning as a method for learning among heterogeneous groups has been used for several decades. Its merits have gained in popularity as a means of eliminating tracking and ability grouping. Implications for cooperative learning and the mathematically gifted student are discussed in Chapter 3.

FAST-PACED CLASSES

The fourth option for mathematically gifted students is studying advanced material at a faster pace than is permitted in the typical classroom. Individually-paced programs ensure that the educational opportunities provided for the children are tailored to their individual needs by placing them at the correct level, presenting material at an appropriate pace, and following a specific, sequential plan. The disadvantages to this type of program include resistance by elementary school personnel, difficulties with scheduling and credit issues, and insuring continuity from year to year (Daniel & Cox, 1988).

One example of a program that successfully combined problem solving and acceleration is the Cumberland Accelerated Mathematics Project (Wheatley, 1988). This project was designed as a fifth and sixth grade mathematics program for gifted students. Its unique features included the following: (a) an emphasis on problem solving using computers and calculators; (b) identification of weak computational skills via diagnostic testing; and (c) absence of arithmetic as an instructional objective. Flexible pacing, as conceptualized by Daniel and Cox (1988), permits able learners to move through the curriculum as their needs, abilities, and motivation require. Students are encouraged to take fast-paced courses or participate in individually-paced programs. An example of an individually-paced program is the Diagnostic Testing → Prescriptive Instruction (DT→PI) model in which instruction is based on the results of diagnostic testing and takes into account relative strengths and weaknesses of the individual student. The DT→PI model is described in Chapter 4.

CURRICULUM AND MATERIALS

When searching for resources that will help them plan their talented youngster's program, parents and teachers report experiencing difficulty in finding appropriate materials. For example, one parent said, "[My 8-year-old child] was recently formally tested in school and qualified for the 'gifted' program, but there is none in our school for math. The various teachers and principal do not quite know what to do with her at this point." Talented students need challenging materials presented at a pace that will provide them the opportunity to develop a thorough background in general mathematics and allow them to progress to more advanced mathematics when they are ready. These students need resources in the form of well-prepared teachers, excellent textbooks, computer software, and other materials. The new materials can be used within and around the traditional sequence. Materials recommended for mathematically talented youth are included in Chapter 5, and other helpful resources are listed in Chapter 8 and Appendix D.

CASE STUDIES

In order to illustrate how the materials and programs we have described can be used, in Chapter 6 we have included case studies of seven children. They range from a talented 10-year-old whose parents did not advocate for him to seek greater academic challenges to a 10-year-old who is involved in a school/home conflict concerning his mathematics education and a 9-year-old who may be ready to study geometry. Although all of these children are talented, their learning needs vary considerably.

ASSESSING CHILDREN'S EDUCATIONAL NEEDS

Children demonstrating intellectual characteristics that are more advanced than their peers are often referred to a psychologist for an academic assessment. This assessment usually includes an individualized intelligence test. Although an intelligence test gives information that is generally useful, it is not designed to determine the extent of mathematical talent. Therefore, intelligence tests alone cannot provide the detailed information needed for making an educational plan in a specific subject such as mathematics. Whereas intelligence tests are a valuable measure of general ability, achievement tests are designed to measure what has been learned. Achievement testing helps the student, parent, and educator determine what needs to be learned and how best to learn it. Testing is never meant to be an end in itself, but only a means to attaining the best approach possible for the student's learning.

Baroody (1987) states that, "To build a theory about a child's mathematical knowledge as completely and accurately as possible, diagnostic efforts should collect data on informal knowledge, specific strengths and weaknesses, skill accuracy and efficiency, concepts, strategies, and errors" (p. 62). For students who are disadvantaged or physically handicapped, an individualized assessment is critical as traditional group assessments may not identify them. In Chapter 7 we give guidelines on what to expect from an educational assessment, and how that information can be used in programming.

PLANNING AND RESOURCES

In addition to the needs specific to mathematics, children need appropriate experiences in other areas. Talented children need to be challenged by mathematics, but do not require math as the major activity in their formative years. They need a balance of activities in all areas. For example, 7-year-old Pam is working at the fifth grade level in mathematics and meets with her mathematics mentor once a week for two hours. This leaves plenty of time for attending dance classes, reading mysteries written for third graders, and talking on the phone with her best friend.

When developing a program for mathematically talented students, it is necessary to think not only of the child's immediate academic and social needs, but also how their needs can be met throughout their school years. In Chapter 8 and Appendix D, we describe programs, materials, and resources designed to aid parents and educators in developing long-term plans.

APPENDICES, GLOSSARY, AND INDEX

To further aid our readers, we have included four appendices: (a) forms to aid educators and parents in implementing mathematics programs for their talented students; (b) problems and solutions from the Mathematical Olympiads for Elementary Schools (MOES); (c) a description of the constructivist's developmental theories of mathematical talent; and (d) an extensive list of programs and resources. A glossary of educational terminology and an index conclude the book.

HISTORICAL AND CURRENT PERSPECTIVES

Our philosophy toward educating mathematically talented students is rooted in a tradition that was established in exemplary fashion by Leta Hollingworth over half a century ago. Anyone who reads Hollingworth's (1942) classic, *Children Above 180 IQ Stanford Binet: Origin and Development*, cannot help but be struck by the similarity of the educational concerns of the 1920's and '30's and those expressed by today's educators of academically talented students. Statements such as the following, made in 1939, are as true today as they were then.

> ... the problems [of education of gifted elementary pupils] are most urgent on this level, because it is in the primary and elementary school that the very intelligent child most especially needs a supplement to the standard curriculum. The program of progress through the elementary grades is based on what pupils at, or only very slightly above, the average can master at given ages, so that the extremely intelligent child has little or nothing to do there. . . ." (pp. 307-308)

How shall a democracy educate the most educable? At present these children are to a great extent lost in the vast enterprises of mass education, and are left to handle their special problems as they may, by themselves, while the energies of teachers are bent upon the main business of dealing with the [other] ninety-nine per cent who test below 130 IQ [Stanford-Binet]. . . . It is for us to consider [academically gifted children] carefully, for educators are the sole group appointed by society to guard the interests of children. We are their official

History

guardians, adding our guidance to that of their natural guardians, parents, who are often helpless either to recognize these children's abilities or to develop them. (p. 318)

How were these academically gifted students, for whose education Leta Hollingworth was so eloquently concerned, educated? In January 1936, the Board of Education of Manhattan, NY and Teachers College of Columbia University founded the Terman Classes at the Speyer School. The goal was to provide a curriculum that would "extend their minds and interest them in the interests of society during the years of the elementary school" (p. 309). What can we learn from the experiences of the Terman Classes that will aid us in formulating the best educational plan for today's youngsters?

Students were selected according to age (from 7 years, 0 months to 9 years, 6 months), representativeness of the variety of ethnic groups then living in New York, and IQ (130 or above on the Stanford-Binet). Hollingworth described a curriculum that was advanced as well as enriched. The professionals of her era had determined that the student with an IQ of 130 could *master* the elementary curriculum in half the time needed by the average student. This was the advanced aspect of the Speyer school curriculum. The enriched aspect featured a foreign language, literature, and a series of units falling under the general classification of "The Evolution of Common Things." Thus we see a concern for balance as reflected in the selection procedures and the curriculum. However, it is disconcerting to think that issues which were seen so clearly by Hollingworth and her colleagues over 50 years ago are still of concern, and Hollingworth's sound approach is not the accepted norm for educating today's extremely talented students!

Hollingworth recognized that although an individual intelligence test score provided an important piece of information, it was not adequate for establishing a program plan. Hollingworth used the intelligence test as the first measured indication of a student's potential. To further delineate a student's talents and aptitudes, she administered more specialized tests. Her philosophy was simple: use the tool that most accurately measures the strengths of the student. For extremely talented students, the typical age-in-grade level test was not a precision instrument because students tended to top out on them by answering all, or nearly all, of the questions correctly. Instead, she used tests developed for older students. Thus, we have an early model for our work with mathematically talented elementary students.

Hollingworth's work was prematurely terminated by her untimely death in 1939. Nevertheless, her influence has been sustained in the work of Julian C. Stanley's Study of Mathematically Precocious Youth (SMPY), which is described below.

STUDY OF MATHEMATICALLY PRECOCIOUS YOUTH

The philosophy underlying SMPY is set forth by Stanley and Benbow (1986). Briefly, SMPY identifies mathematically talented students by giving them a test developed for much older students. Talented 12-year-olds take the Scholastic Aptitude Test (SAT), which was developed for college-bound high school juniors and seniors. Then, high-scoring students are encouraged to participate in a variety of advanced classes and mentor-paced instruction, where they study material at a pace faster than is typical but matched to their more advanced abilities. Stanley's work with the SMPY model provided the foundation for other programs that are quite well-known throughout the country such as Johns Hopkins University's Center for the Advancement of Academically Talented Youth, Duke University's Talent Identification Program (TIP), Northwestern's Midwest Talent Search, and the University of Minnesota's Talented Youth Mathematics Program. These are discussed in Chapter 8 and Appendix D.

When he founded SMPY in 1971, Stanley chose to focus on 7th grade students because the first year of algebra causes serious motivational problems for mathematically talented youngsters. No matter how quickly these students might be able to learn algebra, they were lock-stepped into the traditional pace. The only exceptions at the time were a few schools that permitted 8th graders to take 9th grade algebra. The SMPY staff recognized that some elementary-aged students needed mathematically challenging coursework, but chose to begin their work by focusing on 7th grade students (Stanley & Benbow, 1986). Over the years individuals associated with SMPY have developed programs and curricula for young mathematically talented students (Benbow, 1991; Cohn, 1988; Cohn, 1991; Daniel & Cox, 1988; Lupkowski, Assouline, & Stanley, 1990). The work with junior high students has provided us with a sound base for our work with mathematically talented students younger than age 12.

GIFTEDNESS AND MATHEMATICAL TALENT

We do not have a simple definition for a mathematically talented student. However, four theoretical approaches to talent development provide a useful framework from which we have built our model for educating mathematically talented elementary students. The **psychometric approach**, characterized by standardized tests and scores, is probably the most prevalent approach. Three other approaches: the **cognitive developmental approach, information processing**, and **multiple intelligences** also play an important role in helping educators understand mathematically talented youngsters.

PSYCHOMETRIC APPROACH

The psychometric approach aids us in identifying students and in determining where to begin instruction. In many school districts, gifted students are identified primarily on the basis of intelligence test scores. Implicit in the psychometric approach are the assumptions that traits measured by standardized tests vary from person to person and that it is informative to compare one person's performance to that of a group. The gifted individual has a vastly superior performance on tests of mental ability when compared to the average individual of a group. In psychometric terms this is described as scoring several standard deviations above the average. An IQ of 130, for example, is two standard deviations above the average IQ of 100. (See Chapter 7 for a more thorough explanation of this concept.) Much research has been done from a psychometric perspective, beginning in full force with Terman in the early 1900's and continuing to this day.

COGNITIVE DEVELOPMENTAL APPROACH

The cognitive developmental approach assists educators in understanding underlying cognitive processes, and how these processes are developmentally dependent--both chronologically and mentally. This information then clues the educator into when instruction of certain concepts should begin. The cognitive developmental approach, as theorized by Piaget, identifies universal structures of human thought and the operational transformations that bring about these structures. Cognitive developmentalists describe human development in terms of an invariant, universal hierarchy of stages. For our purposes, knowledge of Piagetian stages is very useful in understanding the development of mathematical reasoning and in explaining some of the apparent inconsistencies that young students demonstrate in their understanding of elementary mathematics.

Little has been done empirically to link the cognitive developmental approach with the field of gifted education. Feldman (1982) asserts that, because developmentalists try to discover the common processes of change, they have not been concerned with applying developmental theories to the field of gifted education. Nevertheless, two important studies yielded results that have practical significance for our work with mathematically talented youngsters. In the first study, Keating (1975) examined giftedness as defined psychometrically as well as developmentally in Piagetian stage theory. In brief, he found that bright 5th graders, as identified by the *Raven's Progressive Matrices* (a test of nonverbal abstract reasoning ability) were at a more advanced cognitive developmental level than average seventh grade students. The practical implication of this finding concerns placement in curriculum: *what is appropriate for the average seventh grader is not appropriate for the precocious fifth grader.*

A second study, conducted by Carter and Ormrod (1982), tried to sort out the equivocal results of investigations concerning the relationship between chronological age, mental age, and the acquisition of concrete and formal operational thinking. They found that children with high academic ability progressed more quickly within a stage. The gifted children in their study also demonstrated earlier transition to the next stage. The educational implications of these results are similar to those from Keating's (1975) study.

INFORMATION-PROCESSING APPROACH

Information processing as a psychological movement emerged strongly in the 1960's and 1970's (Gallagher & Courtright, 1986). This approach describes how individuals store, retrieve, and use information. Its emphasis on questioning individuals to determine how they solve problems adds an important dimension to the process of identification and the development of instructional objectives: teachers can interview students to ascertain how they learn and which underlying concepts need to be further developed, then provide enriching experiences so that strengths, e.g., problem solving and computation, are developed concurrently.

Information-processing via Sternberg's triarchic theory of intelligence provides us with terminology for appreciating the mathematically talented student's extraordinary ability. Sternberg's (1986) theory of intelligence is triarchic because it describes intelligence from three perspectives: (a) the internal world of the individual, (b) the individual's experiences with tasks or situations most requiring the use of intelligence, and (c) the external world of the individual. Sternberg has applied his theory to intellectual giftedness in an effort to broaden our thinking with respect to intelligence. Sternberg's theory helps us to see mathematically talented individuals as more than computational wizards and to appreciate the implications for their talent to solve problems in novel fashions.

MULTIPLE INTELLIGENCES

Gardner's (1983) theory of multiple intelligences is a reaction to the psychometric approach, where gifted individuals are often identified based upon their general intelligence (known as the "g" factor) instead of abilities in specific areas. He views intelligence as composed of multiple components, and has identified seven: linguistic, logical/mathematical, spatial, musical, bodily-kinesthetic, interpersonal, and intrapersonal. Identification procedures that emphasize a general intelligence score may exclude students who are talented in a particular area. For example, a student who is talented in the logical/mathematical area will not be able to demonstrate fully that talent on an intelligence test because the test is too general. "One implication of Gardner's theory for identification of gifted students is that the elements or processing strategies associated with each intelligence need to be related to identifying giftedness in young children and then providing ways for the special enrichment needs of these students to be related to curriculum goals and objectives throughout the school years" (Klausmeier, 1986, p. 216).

CONCLUSION

In our book, we have tried to present a variety of perspectives regarding educating mathematically talented elementary students. The title of the book, *Jane and Johnny Love Math*. . . reflects the opening statement by the hundreds of parents and teachers who have sought our advice: "My child loves math, but nobody knows what to do with him [or her];" or, "I have a student who seems driven to learn more math . . . what do you recommend?"

We describe a variety of students, all of whom are gifted in mathematics, and present several approaches to working with those students. In discussing our views about testing and curriculum and how they relate to mathematical talent, we convey the following points:

* The children described in this book have unique characteristics and academic strengths requiring individualized programming.

* Relevant, objective, and complete information is needed to develop a systematic, challenging program based upon a student's individual needs.

* An educational program that is appropriately challenging will enhance learning rather than stifle it.

History

# MAKING INFORMED EDUCATIONAL DECISIONS

HISTORICAL BACKGROUND FOR AGE-IN-GRADE GROUPING

An eight-year-old child living during the 19th century would probably have attended a one-room school with children ranging in age from five to sixteen. If that eight-year-old were more advanced than his or her age mates, the teacher would have prepared lessons and presented materials according to the student's ability, even if it meant using books intended for children much older.

In contrast to 19th century one-room schools, today's schools are part of a sophisticated educational system that developed from the educational reform movement led by Horace Mann in the mid-1800's. One result of this educational reform was grouping students into grades based on age. Grouping children by age was quite compatible with the industrialization of our country and the Americanization of immigrants. In Joseph Kett's report to the President's Science Advisory Committee, it was concluded that,

> . . . based upon well-intentioned concern for educational reform, the practice [of age-grade grouping] initially was well suited to the political and economic needs of the nation; however, age-grade grouping survives to this day essentially unchanged. (Daurio, 1979, p. 14)

Today, classrooms are filled with children of the same age whose needs vary considerably. Age-in-grade grouping may lead to problems that are especially evident in mathematics classrooms because of the popularity of the spiral approach to studying mathematics. In the spiral approach, material is introduced in the primary grades and then is repeated with minor variations throughout the elementary curriculum. For example, in one textbook series, first graders add two-digit numbers; second-graders add three-digit numbers; and third graders add four-digit numbers. Even if youngsters are ready for more advanced material, they are required to plod along and complete tasks involving skills they may have mastered even before the first grade. As a reflection of the current sophisticated educational system, students are grouped according to age, and materials are essentially the same for students of varying abilities. Making adjustments in the standard curriculum for talented students is often viewed as a problem because it contributes to the teacher's already heavy load, interferes with scheduling, and sometimes requires placement outside of the age group. Differentiating the regular curriculum is necessary for talented students, however, because their learning needs differ from those of typical students.

WHAT IS DIFFERENTIATION?

By "differentiating the curriculum" we mean allowing for individual differences and planning and presenting material according to individual levels of performance. For example, in a classroom where most students are the same age but have varying abilities, teachers might place students into small groups based upon proficiency. This practice maximizes the teacher's effectiveness as well as the students' learning experiences.

In gifted education, "differentiated curriculum" refers to a medley of educational programs including acceleration and enrichment. By acceleration, we mean entering kindergarten or first grade early, skipping an entire grade, or advancing in a subject area more rapidly than is typical. Enrichment programs may include after-school activities that focus on problem solving, in-school opportunities to study subjects not covered in the typical curriculum, or in-depth exploration of topics covered in the regular curriculum. In our view, acceleration and enrichment are not mutually exclusive. Unfortunately, those implementing educational policy have defined acceleration and enrichment as opposites and have often forced educators into comparing the two and declaring allegiance to one side or the other. However, "there is a fundamental difficulty with comparing acceleration and enrichment because the terms do not seem to represent points along the same continuum. Acceleration usually is considered an administrative arrangement; enrichment, as typically defined, refers to a curriculum modification or addition" (Howley, Howley, & Pendarvis,

1986, p. 168). Instead of taking sides in this seemingly unending debate, we favor programs that provide an appropriate match between a child's abilities and the school curriculum.

In order to help parents and educators make informed choices for educating mathematically talented youngsters, we have listed a number of options that are available. These options include maintaining the status quo, enriching the standard curriculum, accelerating students, and developing an individualized program of study. No single option is the "best" for all students; the option of choice for one student may even be detrimental to another student's progress. The abilities of the student, his or her social and emotional maturity, educational resources, and other local circumstances must be taken into account when deciding how best to meet the needs of that particular individual.

WHAT ARE THE OPTIONS?

1. MAINTAIN THE STATUS QUO

Some students have demonstrated the ability to handle more advanced material, but seem quite satisfied with the current educational program and do not want to see it changed. This is one way of opting to maintain the status quo. Whether or not to encourage these students to pursue more challenging coursework is a decision that must be determined carefully by the student, parents, and educators.

Some who favor maintaining the status quo argue that making exceptions or developing special programs for talented students is unfair, and that these students should be expected to do the same work at about the same pace as other students in the class. This argument is rooted in a founding belief of our democratic society, that we are all equal. This belief has been interpreted to mean that we should provide the same education for all students. However, as Thomas Jefferson said, "There is nothing more unequal than equal treatment of unequal people." Although we believe in providing equal *opportunities* for learning to all children, we recognize that the learning needs of students vary considerably, necessitating curricular adaptations.

When required to sit and listen day after day to material already known, a child may respond in a number of ways. "He or she can daydream, be excessively meticulous in order to get perfect grades, harass the teacher, show off knowledge arrogantly in the class, or be truant. There is, however, no *suitable* way to while away the class hours when one already knows much of the material and can learn the rest almost instantaneously as it is first presented. Boredom, frustration and habits of gross inattention are almost sure to result" (Stanley & Benbow, 1986, p. 368). Bored students are at risk for behavior problems and losing interest in a subject that previously brought much pleasure. If a program is developed that is based upon individual needs, however, the child receives an appropriately challenging education and cannot be accused of receiving special privileges.

2. COOPERATIVE LEARNING

In cooperative learning, small teams of students with a range of abilities are grouped together to promote peer interaction and cooperation while learning academic subjects. Much support for cooperative learning is generated from the democratic ideal of our society that those who have more should help those who have less. Cooperative learning has gained in momentum as concerns about tracking and underrepresentation of minorities in gifted programs have increased. Cooperative learning has also been viewed as a solution to concerns educators have regarding the affective development of their gifted students. Cooperative learning can be an effective technique. However, the way it is often used with gifted students exploits their ability.

Careful examination of the research indicates that cooperative learning often is effective with average students. It has also been found to be effective when gifted students are grouped together. However, when the cooperative groups are composed of students with diverse abilities, the gifted student frequently becomes the tutor for his or her peers. Oftentimes the tasks required of the group are not challenging for the gifted students. In order for cooperative learning to be effective with gifted students it is recommended that:

(a) gifted students be grouped together occasionally or allowed to choose their own groups,
(b) when group members take on different roles, gifted students have roles that require higher-order thinking,
(c) projects and assignments are created that do not allow groups to depend on one person's efforts, and
(d) group grades be avoided (ASCD Update, 1990).

3. COMPLETE SUPPLEMENTARY TASKS AND ASSIGNMENTS

Some teachers ask talented students to tutor their less-advanced classmates. Although peer assistance may provide valuable opportunities for social interaction, we feel that it is neither the appropriate role nor an appropriate educational goal for either student. In addition to keeping talented students from moving ahead at a suitable pace in mathematics, peer tutoring may be quite harmful for less-advanced students, as it may diminish their self-concepts. Few children have the maturity needed to serve as tutors for others. In general, talented students have a different, more rapid way of solving problems and may not be able to explain their thinking to others.

In some cases, when talented students consistently finish their work before the rest of the class, teachers give them additional work to keep them occupied. For example, bright students may be assigned 40 multiplication problems instead of the 20 given to the other students. After completing their regular assignments, talented students may be required to do additional worksheets while their classmates finish their seat-work. Although these activities may keep the students busy, the extra work may have the undesirable effect of boring the student and making him or her careless.

In other programs, talented students may be doing more advanced work than their classmates, but also are required to complete the same assignments as their classmates. For example, a first grader who is studying fourth grade math may be required to complete the first grade mathematics curriculum concurrently. This requirement comes from the legitimate concern of educators that students might finish a grade with gaps in their basic skills. However, if a young student can master the fourth grade content of a sequential subject such as mathematics, teachers and parents can be confident that there are no "holes" in the child's skills. "Some educators are overly concerned with possible discontinuities in instruction caused by differentiating a program and ignore the true omissions that occur when the gifted are denied opportunities for more advanced learning" (Howley, Howley, & Pendarvis, 1986, p. 142).

4. ENGAGE IN ENRICHMENT ACTIVITIES

One means of differentiating the curriculum is to provide enrichment activities. "Enrichment may mean the addition of an extra subject in the curriculum, extra trips to special-interest locations, additional topics covered in the individual curriculum areas, or more homework. Ideally, it means that the pupil's education will be broader in scope, exploring topics in greater depth and at higher levels of difficulty and involving many activities not covered in the regular program" (Ehrlich, 1985, p. 13).

Curriculum compacting (Renzulli & Reis, 1991; Renzulli, Reis, & Smith, 1981) is a well-developed method for eliminating unnecessary repetition of material already learned. The goal of curriculum compacting is to determine what the student knows and eliminate that material from required lessons. Curriculum compacting is appropriate for above average, bright students. However, for the *extraordinarily* talented student, curriculum compacting does not provide a systematic approach to identify the *specific learning needs* of the student.

Enrichment activities unrelated to mathematics.

The typical enrichment program is designed so that students remain in their grades and have the opportunity to study material unrelated to the classroom curriculum. Enrichment material is frequently taught on a weekly basis by a special teacher and is often implemented by a school system because of a recognized need to do something "more" for gifted children.

Although such enrichment programs may provide the student with fun and stimulating activities, the activities are generally irrelevant to the specific academic needs of the student. When these programs were designed, the specific learning needs of the students were not considered. In fact, these enrichment programs sometimes have "deceleration" as their goal. In other words, the student is held back or slowed down from learning material for which he or she is ready. These programs do nothing to relieve the boredom experienced by the student during the typical school day, other than perhaps to provide a temporary respite from it.

Problem solving activities.

Problem solving has become a popular component of enrichment programs. In some schools, the "Gifted Program" is comprised entirely of problem solving materials and activities. With the flood of problem solving materials and activities on the market, it is tempting to rely upon problem solving worksheets for enrichment of the regular curriculum. For example, at one elementary school, talented students remain within the regular mathematics classroom, complete the regular work, and use problem solving activities as enrichment.

Problem solving in specific areas, especially mathematics, has become a major portion of the mathematics curriculum. The *Curriculum and Evaluation Standards for School Mathematics* (1989) developed by the National Council of Teachers of Mathematics (NCTM) lists "Mathematics as Problem Solving" as the first standard for grades K-12. When referring to the K-4 curriculum, it is stated that "Problem solving should be the central focus of the mathematics curriculum" (p. 23). We agree whole-heartedly with the importance of learning and practicing problem solving skills. However, we do not view problem solving as the only means for differentiating the curriculum for extremely mathematically talented students. As the NCTM states, problem solving is a primary goal in mathematics education for *all* students. *When a curriculum is appropriate for all students, it is not differentiated instruction.*

Mathematically-oriented enrichment activities.

In a publication concerning programming for the 1980's, the National Council of Teachers of Mathematics (NCTM) set forth their position regarding educating gifted children:

> In general, programs for the gifted should be based on a sequential program of enrichment through more ingenious problem solving opportunities rather than through acceleration alone. (NCTM, 1980, p. 23)

In a later publication designed specifically to address programming for children gifted in mathematics the NCTM asserted that:

> The needs of mathematically talented and gifted students cannot be met by programs of study that only accelerate these students through the standard school curriculum, nor can they be met by programs that allow students to terminate their study of mathematics before their graduation from high school. . . . (House, 1987, p. 100)

Belcastro (1990) called on NCTM to withdraw its position regarding acceleration because it is not founded upon research evidence. Teachers who make recommendations based upon these position statements risk denying talented students the opportunity to learn appropriately challenging material.

Some rare programs, which have been labeled as enrichment, have been carefully designed based upon the needs of the students (Stanley, 1979b). These programs are intended to challenge students with relevant materials and thereby relieve their frustration and boredom. Enrichment programs that do not decelerate or hold the student back will begin to take on accelerative characteristics. As noted by Anastasi (1979), ". . . the more closely the content of enrichment matches individual interests and talents, the closer it approaches acceleration" (p. 222). The issue is not the label on the program but the match between the student's needs and materials provided by the program. Good enrichment programs will have a fair amount of acceleration, and good acceleration programs will have a fair amount of enrichment.

Examples of mathematically-oriented enrichment activities include contests such as the Mathematical Olympiads for Elementary Schools (MOES), computer games, and recreational mathematics books. More information about these activities can be found in Chapter 5 and Appendices B and D. Computer games, and recreational math books may be especially appropriate for the above-average mathematics student and for the extraordinarily talented primary (preschool through grade 3) child. For example, Mark, an extremely talented 5-year-old, had nearly completed the 8th grade mathematics curriculum in a home-schooling situation before his parents contacted us for advice. We were concerned that he wasn't quite ready for algebra and advised the parents to proceed with caution and to use mathematically-oriented enrichment activities to supplement his learning.

5. ACCELERATE THE EDUCATIONAL PROGRAM

Students can choose from a "smorgasbord" of accelerative opportunities. Entering kindergarten early, skipping a grade, skipping ahead in one subject, taking fast-paced classes, condensing four years of school into three, receiving credit by examination, and working with a mentor in an individually-paced program are among the 12 educational options listed by Benbow (1986).

Entering kindergarten early.

Pre-kindergarten screening is widely available in school districts throughout the United States and is intended to identify children with special needs. This screening may be the first time that children's abilities are assessed and that parents are made aware of their child's specific talents. Extremely talented children might be given the option of entering kindergarten at the age of four instead of five. In order to participate effectively in kindergarten activities, it is critical that the preschooler have not only advanced intellectual abilities, but also the necessary motor and social skills. It is also important to investigate the kindergarten program before placing the child to ensure that the curriculum is sufficiently challenging and that needed adjustments can be made.

Grade-skipping.

Children who are evenly advanced in the various academic subjects and who have demonstrated social maturity are good candidates for skipping one or more grades. In order to make the transition smoother, it makes sense to take advantage of the natural breaks in the educational cycle, such as semester breaks and the end of the school year. One of the easier times to skip a grade is when a building change is involved for the entering grade. For example, when going from elementary to middle school, a child might skip the last year of elementary school, thus entering middle school a year early. If there is an older sibling in the same grade that the advanced child would be entering, educators and parents need to plan carefully because of the possible negative effects on both siblings.

Even though these children may have fairly evenly advanced development in all skill areas, skipping a grade may not be the ideal way to challenge them intellectually. The pace at which material is studied one grade higher may still be too slow, as their classmates may be less cognitively mature than the extremely able child. For example, a young boy who was advanced in all areas skipped from first grade to third grade. He was also extremely talented in mathematics and therefore was placed in a regular high school geometry class. Although he was studying material that was quite advanced considering his age, he was bored because the pace was still too slow and completed the course on his own.

Fast-paced classes.

There are a few programs in the United States that offer talented elementary students the opportunity to study mathematics at a faster pace than is typical. One example of such a program is the Young Students Classes developed by the Center for the Advancement of Academically Talented Youth (CTY) at Johns Hopkins University. Classes for elementary students meet after school and on weekends,

Making Informed Decisions

permitting talented students to study mathematics with talented peers. CTY staff work with students, parents, and school officials to ensure that participating students earn school credit for the work they complete and are placed in the appropriate classes when finished with the Young Students Classes (Moore & Wood, 1988). Similar programs are offered by Arizona State, Carnegie Mellon, Sacramento State, University of North Texas, and The University of Iowa.

Condensing four years of school into three.

Students may be permitted to shorten a program by condensing four years of school into three. For example, one advanced student was promoted to 4th grade after spending only three years in a K-3 program. Other students might complete the work for both first and second grades in one year, and thereby be eligible to move on to third grade earlier than is typical. This method of acceleration is more common at the high school level.

Credit by examination.

Some students learn a great deal of mathematics outside of school by working with their parents, reading books, and exploring math on their own. To determine the level of achievement, school personnel might administer end-of-the-year standardized tests. This may result in receiving credit by examination. For example, one fourth grader took the fifth grade final examination and did well enough on it to be promoted to the sixth grade math class.

Subject-matter acceleration.

Children who are extremely talented in one subject may not have equal strengths in all other areas. In that case, it may be in the child's best interests to allow him or her to study one subject with older children while remaining with his or her age peers for other subjects. Often, this is accomplished by having the student attend the math class one or two grades higher. However, concerns about the slow pace of the class raised when discussing grade-skipping are equally valid for subject-matter acceleration.

Another method of subject matter acceleration, often used because of scheduling conflicts, involves having students work alone on advanced materials. We caution educators and parents not to fall into this trap of having students work through the textbook on their own. Although it might be an easy way to "challenge" students, it is not the ideal method for learning mathematics. Rushing through mathematics books should not be the primary goal for extremely talented elementary-age students. So that these students may progress at their maximum ability level, they need sustained adult guidance, such as that provided by mentor-paced learning as described in Chapter 4.

Individually-paced programs.

A final accelerative option is an individually-paced program. If a talented student is permitted to work with a mentor at an appropriate pace, the student will move more quickly through the curriculum than is typical. A complete explanation of an individually-paced program, SMPY's Diagnostic Testing → Prescriptive Instruction model, is provided in Chapter 4.

COMMENTS CONCERNING ACCELERATION

SOCIAL DEVELOPMENT

Social development is the major consideration in the decision-making process regarding acceleration. Opponents of acceleration are outspoken, mainly in terms of the perceived negative effects on the social and emotional development of the accelerated child. Opposition is often based upon an experience that school personnel and parents may have had with one accelerated child who, for various reasons, had problems making the adjustment. However, research has repeatedly shown that acceleration does not necessarily result in social maladjustment (for reviews of this topic, see Daurio, 1979; Janos & Robinson, 1985; Kulik & Kulik, 1984, 1987). Assouline, Colangelo, and Lupkowski (1992) have developed an acceleration scale, *The Iowa Acceleration Scale*, that is intended for use as a guidance tool in making decisions about advancing students one or more grades.

It is possible that accelerated students will have fewer opportunities to interact with students of the same age. However, we do not view this as a sufficient reason to hold a student back. Given the importance of learning how to live and get along with people of all ages, fewer opportunities to play and work with agemates does not seem to be a convincing argument against acceleration. There are usually plenty of chances for peer interaction, either in informal play with neighborhood children or during organized activities such as scouts, band, sports, etc. What is important is the opportunity to form friendships with others who have the same interests. Having many acquaintances who share a birth-year is not as critical as having a few close friends with whom to share important ideas.

A helpful indicator in evaluating the social readiness of a child to be accelerated is to consider the ages of favorite playmates. For example, when Jane was four years old, she enjoyed playing games with rules. Since the other four-year-olds in the neighborhood weren't interested in those games, she played with six- and seven-year-olds who lived nearby.

Another concern is founded in the likelihood that students who skip one or more grades may encounter social situations for which they are not yet ready. "Acceleration of academic placement (grade skipping) can often improve the match between capability and intellectual challenge and provide exposure to a wider range of non-gifted young people, but it may throw the gifted youngster into a social setting demanding more maturity than he or she has achieved" (Robinson & Robinson, 1982, pp. 79-80). That problem may be intensified during the adolescent years.

Although the social needs of an eight-year-old may not differ drastically from those of ten-year-old classmates, six years later, the young fourteen-year-old student will have fewer social privileges (such as driving and dating) than sixteen-year-old classmates. Although most accelerated students say that the age differences aren't a major problem, they do acknowledge some minor inconveniences that disappear in a few years. The students say those minor inconveniences are well worth the privilege of getting more advanced coursework and not having their learning restricted (see Brody, Lupkowski, & Stanley, 1988).

In this discussion, we have talked about five important points:

(1) Healthy social development means learning to get along with people of all ages and skills.
(2) There are many opportunities outside of school to interact with age peers.
(3) Having a large number of same-age friends isn't as critical to a child's healthy social development as having a few close friends with whom to share ideas.
(4) An indicator of social readiness for acceleration is preference for playing with more mature children.
(5) Finally, long- and short-term planning is essential. Students, parents, and teachers need to remember that there may be a few years during adolescence when minor inconveniences become major ones.

A number of books and newsletters have been written on the social development of gifted children (e.g., see Adderholdt-Elliott, 1987; Clark, 1988; Delisle, 1992; Delisle & Galbraith, 1987; Milgram, 1991; and the newsletter, *Understanding Our Gifted*, see Appendix D). Our perspective concerning social development includes insuring opportunities for satisfying interactions with other youths, but our focus is not only on interactions with age-mates. We advocate opportunities for interaction with children of all ages and skills.

SOME ADVICE TO PARENTS

Parents considering acceleration may be accused of "pushing" their children. Well-intentioned parents should be prepared for such an accusation. Most parents consider accelerative strategies in response to their children's needs. Unfortunately, some push their children ahead because they want to break a record or "look good." Encouraging a child to move ahead faster to break a record or to perform "amazing intellectual feats" in order to feed parents' or teachers' egos is categorically wrong because these actions involve exploiting the child.

Over the years a few stories about precocious children who were the victims of pushy, publicity-seeking parents have become well known. One of the most frequently mentioned cases is that of William James Sidis, the precocious youngster with an IQ over 200 who entered Harvard at age 11. William's parents believed they could create a genius. During his youth, newspapers were filled with accounts of his achievements. Instead of shielding him from reporters, William's father advised him how to manipulate them, and even published a book that drew enormous attention to the youngster. By the time he was 16, William was desperate for privacy. He rebelled against his parents, and became more and more of a recluse. William's decline was not a result of intellectual "burn-out;" rather, it was an attempt to escape publicity. He died at the age of 46 in a self-imposed exile (Wallace, 1986). Sadly, the opponents of acceleration often cite stories such as William's, and blame his problems on his acceleration rather than on the mismanagement of his life by his parents.

Precocious youngsters are often featured in the media because their lives make fascinating human interest stories. Features in newspapers and television segments on precocious or prodigious youngsters are not only informative to the general public, but also can offer a child experiences that might not have been available otherwise. Such experiences include contact with professionals who can offer valuable advice and encouragement, as well as special educational and employment opportunities. However, we caution against overdoing the publicity. Herman (1982) lists some considerations for parents who may wish to publicize their child's achievements:

(1) Publicity may make a child vulnerable to ridicule by other children and teachers.
(2) Shy children or very young children may lack the maturity and poise to deal with television cameras and reporters' questions.
(3) Pre-teens should be interviewed only when a parent is present, and teenagers should be permitted to decide whether or not they want the publicity.
(4) Parents and children should be sure to state all facts clearly, avoid saying anything that could be overly negative or a source of embarrassment, and try to discourage reporters from sensationalism.
(5) Parents should discuss the situation with their child before and after the interview, allowing an opportunity to express his or her views.

ADVANTAGES AND BENEFITS OF ACCELERATION

Acceleration accomplishes much that is useful. When implemented properly, it guarantees that an appropriate match between the talented student and curriculum has been found. Acceleration provides the opportunity for challenges that children might not otherwise experience.

If a student is not permitted to accelerate, he or she may be locked into the traditional pace and sequence of study and denied the opportunity to study a more advanced subject later. For example, a student who waits until ninth grade to study algebra with the rest of his or her classmates may not be able to schedule more advanced mathematics courses such as calculus or differential equations, or advanced science courses requiring high level mathematics. Although one might counter that these subjects can be studied in college, we disagree with that argument for the same reason stated above. Specifically, waiting to study a subject eliminates the chance to take new, more challenging courses. The mathematically talented student who craves higher level mathematics and related subjects would be wise to pursue those interests.

Another positive feature of acceleration is the opportunity to save time. Because of the time saved, students can study more subjects and have more time to study topics in depth. This has two long-term benefits that are realized when students enter college. First, students with advanced standing can graduate early and therefore save the cost of a year or more of college. Second, students with advanced standing who choose to remain in college for four years will have the time to take extra courses, to pursue double or triple majors, or to graduate with both bachelor's and master's degrees.

CONCLUSION

Because learning occurs only when there is an appropriate match between the cognitive level at which a child is functioning and the circumstances that he or she encounters (Hunt, 1961), programs for talented students must be carefully developed based on each individual's needs. By carefully assessing the student's level of general ability, specific aptitudes, and achievement and matching the curriculum to them, a satisfying, challenging educational program can be developed. For some students, enrichment programs will provide enough stimulation; for others, accelerated progress through the curriculum will be the only way to be adequately challenged. Parents, educators, and students should weigh all of the available information when making decisions about educational programs.

Making Informed Decisions

4
DIAGNOSTIC TESTING FOLLOWED BY PRESCRIPTIVE INSTRUCTION: SMPY'S DT→PI MODEL

OVERVIEW

Dr. Julian C. Stanley, founder and director of the Study of Mathematically Precocious Youth (SMPY) at Johns Hopkins University, developed a diagnostic/prescriptive model for teaching mathematics to students with extraordinary mathematical aptitude. The foundation of the model is beyond-level testing. Students who score high enough on screening instruments can then participate in beyond-level ability and achievement testing, which is conducted to determine what aspects of a particular topic have not been mastered. Then, students work with mentors until these topics are well understood. The process concludes with a post-test; students who perform satisfactorily on the post-test are ready to re-enter the process at the next level by taking the test for the next topic. Each one of these steps is described in detail in this chapter. In addition, we describe two university programs that use the DT→PI approach and how those mentor-paced programs have interfaced with the schools in the community. Curricular materials and assessment strategies used in this model are described in Chapters 5 and 7.

BEYOND-LEVEL ASSESSMENT

Some students are so exceptionally talented in mathematics that when they take mathematical aptitude and achievement tests developed and normed for their age or grade they answer all or most of the items correctly. Therefore, the evaluation of their talents is severely restricted. This type of restriction is known as a *ceiling effect*, where students figuratively bounce against the ceiling of a test because there are not enough difficult test items. For example, a student's score at the 99th percentile on a grade-level achievement test reveals that this student answered more items correctly than 99 percent of his or her grade-mates and most likely answered most of the items correctly. However, this does not provide adequate information for programming because it does not tell what topics the student still needs to learn.

In order to measure the abilities and achievements of this type of student more accurately, we recommend using the process of beyond-level testing, a concept implemented by Leta Hollingworth in the early 1900's. *Beyond-level testing* means giving a test that was developed and normed for students who are several years older. By taking a beyond-level (also referred to as an out-of-level) test, the student is presented with challenging items that can eventually be used diagnostically. Having enough difficult items on an achievement test avoids the problem of a ceiling effect.

Dr. Stanley built upon the idea of beyond-level testing to formulate his model of diagnostic testing and prescriptive instruction. Some of the beyond-level procedures that SMPY has pioneered for junior high and high school students can be used with younger students who show promise in mathematics. In particular, the adaptation of SMPY's Diagnostic Testing → Prescriptive Instruction model for use with mathematically talented elementary school students is explained in the next section.

DIAGNOSTIC TESTING FOLLOWED BY PRESCRIPTIVE INSTRUCTION

Diagnostic Testing → Prescriptive Instruction (DT→PI) is a beyond-level model originally developed for use with talented junior high and high school students who were ready to learn algebra at a faster rate and more advanced level than is typical. Our focus here is elementary-age students; however, a brief description of the model as used with junior high students is found at the end of this chapter.

The DT→PI method has been adapted for effective use with elementary students who are extraordinarily talented in mathematics. Children earning a score at or above the 95th percentile on any of the mathematics subtests of a standardized, group-administered achievement test such as the *Iowa Tests of Basic Skills* (ITBS) at their own grade level are recommended for initial screening. Those below the 95th percentile might find the out-of-level testing too difficult and frustrating. However, the 95th percentile is not a strict cutoff; it should only be used as a guideline. If an educator or parent has other evidence that warrants further assessment, the student scoring lower than the 95th percentile should take the out-of-level test.

As shown in Figure 4.1, a five-step procedure of standardized testing followed by instruction based on test results is employed. To illustrate the process, eight-year-old Johnny, who was introduced in Chapter 1, is used as an example. A description of the assessment materials follows the explanation of the DT→PI's five-step procedure.

STEP 1: APTITUDE TEST

Initial screening usually occurs via an in-grade achievement test such as the *Iowa Tests of Basic Skills* (ITBS). For example, Johnny, who was in the third grade, earned a total mathematics score at the 98th percentile of third-grade norms on the ITBS; therefore, he was eligible to try a beyond-level aptitude test such as the Lower Level of the *Secondary School Admission Test* (SSAT) or the *School and College Ability Test* (SCAT).

We recommend using the Lower level of the SSAT to identify mathematically talented third, fourth, and fifth graders. Lupkowski and Assouline (1992) have found that as many as 24 percent of the students who participated in an initial screening were determined to need more individualized mathematics programming.

It is highly important to use a difficult enough level of the aptitude test. Based upon almost two decades of research by SMPY and similar groups, the staff of SMPY uses the following rule of thumb (Cohn, 1988) for determining an appropriate starting point:

> Exceptionally able students in grades one through three should be given the level of the test that is two grades beyond their grade placement, and students in grades four through six should be given the test that is three grade levels beyond their grade placement.

After the out-of-level aptitude test is scored, decisions need to be made concerning what happens next. We recommend that students earn a score at the 50th percentile or higher compared to older students (e.g., for a third grader use fifth grade norms) before moving on to Step 2 (Assouline & Lupkowski, 1992). For those talented students scoring below the 50th percentile (when compared to older students), we do not recommend continuing with the DT→PI procedure. Instead, we encourage curricular adjustments such as enrichment and problem solving activities. These ideas will be further discussed in Chapter 5.

Since Johnny was in the third grade, the Quantitative portion of the Lower Level of the SSAT was administered. This beyond-level test was appropriately selected following the rule of thumb described above. Johnny earned a score above the 50th percentile when compared to fifth graders and therefore was ready to proceed to Step 2.

STEP 2: ACHIEVEMENT PRE-TESTING

Whereas the primary goal in Step 1 was to **discover** mathematical aptitude, the purpose of Step 2 is to **measure** mathematical achievement. This is accomplished by administering a standardized mathematics achievement test such as one from the *Sequential Tests of Educational Progress* (STEP) series or the *Educational Records Bureau Comprehensive Testing Program II* (CPT-II).

Having at least two parallel forms of a test is critical to the DT→PI process because the second form will be used at a later stage. The mathematics achievement test is administered under standardized conditions. Students are encouraged to answer all questions, but are instructed to indicate those items about which they are unsure by writing a question mark to the left of the item number on the answer sheet. When the testing time expires, the examiner retrieves the test materials from the student, scores the test, and determines the percentile rank of the student's score. The percentile rank is based upon the norms for the highest grade-group available, as this represents the most rigorous standard. For example, a second grader who correctly answers 32 items on Form 4A of the *STEP Basic Concepts* test would rank at the 61st percentile when compared to first semester sixth graders. If the same score of 32 were compared to first semester third graders, the student would rank at the 96th percentile. Thus, the sixth grade norms provide the more rigorous standard.

Progress to Step 3 in the DT→PI model is determined by the student's percentile rank on this test. Stanley (1979a) recommends that students score at least at the 50th percentile of the relevant national norms for the grade level of the test administered. Students earning a score at or above the 50th percentile are likely to learn the rest of that subject quickly. Students scoring below the 50th percentile should be given the preceding level of that test. If there is not a preceding level, an easier test needs to be administered.

DT→PI

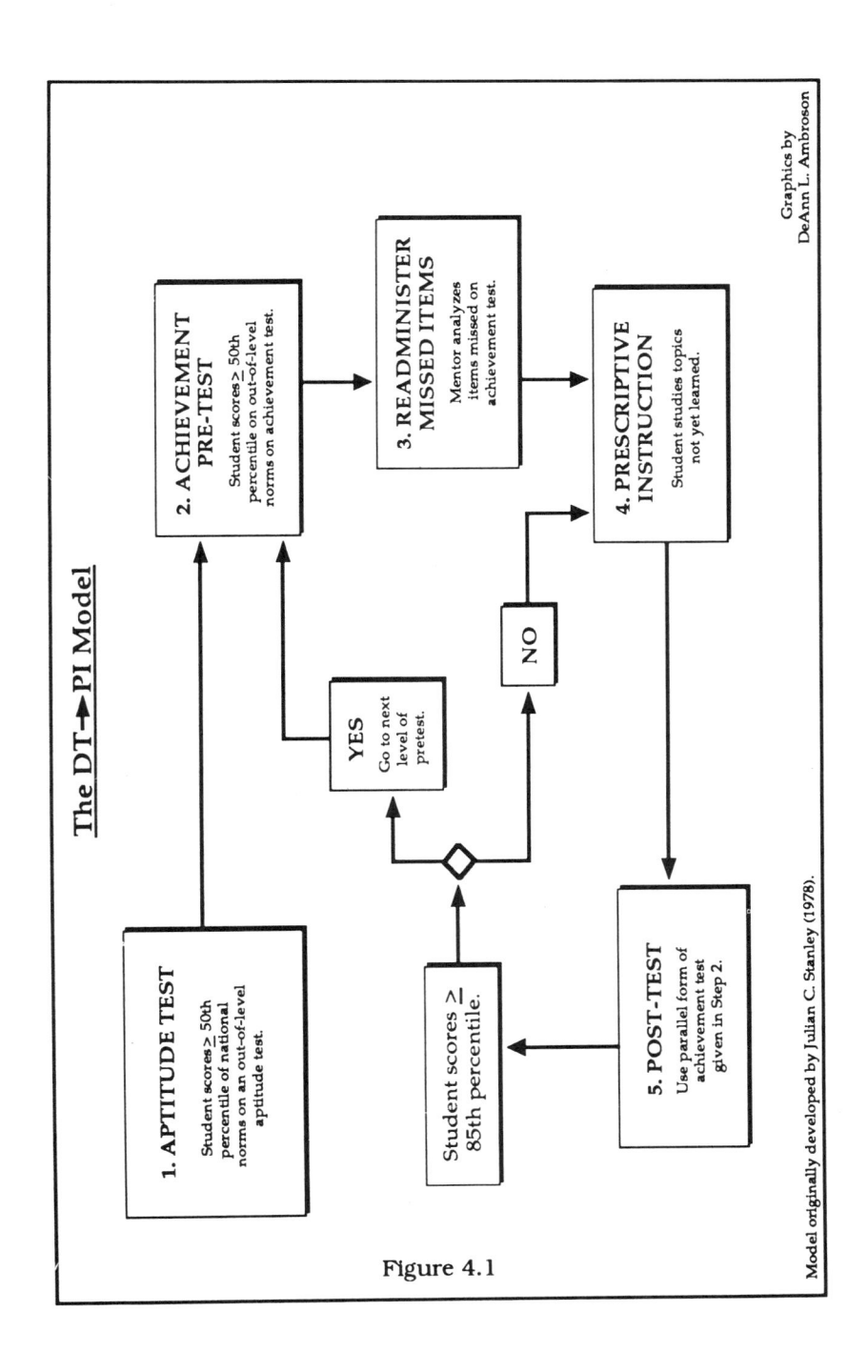

The DT→PI Model

1. APTITUDE TEST
Student scores ≥ 50th percentile of national norms on an out-of-level aptitude test.

2. ACHIEVEMENT PRE-TEST
Student scores ≥ 50th percentile on out-of-level norms on achievement test.

3. READMINISTER MISSED ITEMS
Mentor analyzes items missed on achievement test.

4. PRESCRIPTIVE INSTRUCTION
Student studies topics not yet learned.

5. POST-TEST
Use parallel form of achievement test given in Step 2.

Student scores ≥ 85th percentile.

YES Go to next level of pretest.

NO

Figure 4.1

Model originally developed by Julian C. Stanley (1978).

Graphics by DeAnn L. Ambroson

In Johnny's case, the *STEP Mathematics Basic Concepts* and *Computation* tests, Level 4, Form A (representing a difficulty level appropriate for grades 4, 5, and 6) were administered under standardized conditions. Johnny was instructed to place question marks on the answer sheet next to those items of which he was uncertain. On the Basic Concepts test, Johnny correctly answered 38 of the 50 items, yielding a score at the 83rd percentile when compared to the nationally normed sample of first-semester sixth graders.

STEP 3: READMINISTERING MISSED ITEMS

If the student scores at least at the 50th percentile under standardized conditions, the examiner returns the test booklet to the student with a list of the items he or she marked wrong. The examiner does not return the answer sheet to the student or tell the student how those items were missed. The student is asked to rework those items on a separate piece of paper, taking as much time as needed and showing *all* work. The items are rescored, and instruction is determined based upon the student's responses.

In Johnny's case, he was asked to retry the 12 incorrectly answered items without a time limit and show all work. He correctly answered seven of the 12 items on the second try. Also, he was asked to show the work for those items by which he had placed a question mark during the initial testing. Thus, the mentor learned how Johnny approached the problems and discovered any misunderstandings Johnny had.

STEP 4: PRESCRIPTIVE INSTRUCTION

The fourth step of the DT→PI procedure is prescriptive instruction based on a thorough analysis of the testing results. Prescriptive instruction is conducted by an individual skilled in mathematics, known as a mentor. Although he or she may be the student's classroom teacher, it is not necessary for the mentor to be a trained teacher.

In Step 4, the mentor examines the test items by which the student has placed a question mark, those missed during the initial administration, and the readministered test items. The mentor places the student's responses into one of four categories: (1) items answered correctly, but noted with a question mark, indicating that the student may have guessed; (2) items left blank in the initial administration but answered correctly in the readministration, indicating that time may have been a factor; (3) items missed in the initial administration but answered correctly in the readministration, indicating that the student may have needed extra time (or may have guessed); and (4) items missed in both administrations, indicating that the student did not understand the underlying concept. For items missed twice the mentor needs to examine the student's written solution carefully and determine if the items were missed the same way both times and how the examinee approached the problem.

It is critical to ascertain which items (and their underlying principles) are correctly understood and therefore which other points warrant further instruction. The mentor queries the student about problem solving procedures in an attempt to discover misunderstandings. An item-profile chart (usually available in the test's manual) may aid this process as well as assist in record-keeping. This type of chart can be useful in those programs requiring written instructional objectives as part of the individualized educational plan. Because the categories on this chart are broad, specific objectives should be determined by looking at the concepts measured by individual items.

In addition to using standardized tests to determine a student's achievement in mathematics, it is often helpful to use final examinations and chapter tests provided by textbook publishers as diagnostic and prescriptive pre-testing tools. For example, if results from a standardized test reveal that a student has a weakness on the topic of sets, results from a chapter test on sets would be helpful in uncovering exactly what the student does and does not understand.

The mentor then works with the student on the principles (not the items) he or she does not understand. We recommend using mentor-made problems to help the student study a topic. Textbooks may be useful resources for practice problems. The DT→PI process requires mastering one topic before moving on to the next, but it is imperative that the mentor not require the student to work through every page of a textbook. An advantage of the DT→PI procedure is that students can systematically study concepts in greater depth than typically would be possible and will not needlessly repeat topics already learned. The purpose of the DT→PI model is to help the student learn mathematics well and at a pace that is challenging to him or her. The goal is *not* to race through the elementary mathematics curriculum so that the child might start doing algebra as quickly as possible (Stanley, Lupkowski, & Assouline, 1990). Indeed, the beauty of this process is that it allows the mentor to determine what the mentee does and does not know, and it permits them to work together on the unknown as well as to study mathematics in greater depth. The goal is for the mentee to gain a rich, solid foundation in elementary mathematics that will serve him or her well when studying more advanced mathematics.

Even though Johnny's score at the 83rd percentile comes close to demonstrating mastery of the topics tested, the DT→PI process requires that he thoroughly understand *all* points before moving on. Based upon a detailed analysis of Johnny's performance on the achievement test, the mentor determined what topics needed to be covered. The analysis included an examination of (a) those items by which Johnny placed a question mark; (b) the items he marked incorrectly; and (c) his written solution to the missed problems. These three aspects to the analysis served as the guide to the interview that followed. All of this information was used to determine which principles Johnny did not understand and consequently what content needed to be covered. Steps 2 through 4 were repeated for the *STEP Computation* test, Form 4A.

The mentor worked with Johnny on Saturdays for one to two hours per session. During his school's large-group mathematics instruction, Johnny was permitted to read, work on the computer, or do other homework. Johnny worked on his mentor-assigned homework while his other classmates worked on their teacher-assigned homework. In the evening, Johnny's parents checked that the homework was completed. The teacher was very supportive of Johnny and the completion of the mentor-assigned homework and encouraged him to ask any time he had questions concerning the homework.

STEP 5: POST-TESTING

In the final step of the DT→PI mentor model, the student is given the parallel form of the achievement test as a post-test. The goal is for the student to score *at least* at the 85th percentile of the most rigorous norms for the test, thus indicating mastery of the material. Students who score lower than the 85th percentile require additional instruction and practice with the material. Those who score at the 85th percentile or above, but earn less than perfect scores on the test, require work on the topics they do not yet understand. When the mentor is satisfied that the student has adequately "cleaned up" his or her knowledge of the topic under study, the mentor and student re-enter the model at Step 2, using achievement tests and materials for the next level or topic. Thus, the student studies the mathematics topics in a linear fashion, demonstrating mastery before moving on.

To measure his mastery of the material, Johnny took Form 4B of the *STEP Mathematics Basic Concepts* and *Computation* tests and on each test scored at the 98th percentile when compared to Fall Semester 6th graders. The mentor worked with him on the few items he missed or was unsure of to guarantee mastery of those concepts. Johnny reentered the DT→PI process at Step 2 by taking the next level of the *Mathematics Basic Concepts* and *Computation* tests.

THE ROLE OF THE MENTOR

The role the mentor plays in the DT→PI process is best summarized by Stanley's (1979a) description:

> For the "prescriptive instruction" one needs a skilled mentor. He or she should be intellectually able, fast-minded, and well versed in mathematics considerably beyond the subjects to be learned by the "mentee(s)." This mentor must *not* function didactically as an instructor, pre-digesting the course material for the mentee. Instead, he or she must be a pacer, stimulator, clarifier, and extender. (p. 1)

It is not necessary for the mentor to be a trained mathematics teacher. Engineers, college professors, and undergraduate mathematics majors, as well as high school math teachers have been successful mentors in SMPY and related programs. It is important that the mentor not be threatened by the abilities of the student. The goal of mentoring is to help students learn self-teaching skills, not to spoon feed them information.

If the mentor is someone other than the classroom teacher, it is critical for mentor, teacher, and parents to communicate so that all are aware of the level of instruction presented to the student. For many reasons, parents are usually not good mentors for their own children; they may be too ego-involved or demanding, or may not be able to get enough cooperation from the child, or have credibility with the school.

Conventionally, students sit in a mathematics class for about 45 minutes daily. However, the learning setting in the DT→PI process is more highly concentrated than in the typical classroom, so daily meetings are not necessary. Based upon our experiences, we recommend that a mentor and student meet once weekly for about two hours. In between meetings, the student should work diligently on a considerable amount of mentor-assigned homework. It is helpful if the mentor is accessible to the student via telephone to answer questions or make clarifications. Although the DT→PI process involves mentor-paced, rather than student-paced instruction, it is critical for the "mentee" to take responsibility for his or her own learning, especially by doing homework carefully, completely, and well.

> Homework is an essential aspect of any fast-paced mathematics program. The material is covered in class so quickly that it must be reinforced by extensive homework assignments. The amount of homework should not be so great that it places a burden on the student or becomes repetitive or busy work. Nevertheless, homework can help a gifted youth make the most of his or her abilities. Students . . . must develop good study habits in order to space out the homework over the interval between classes. The discipline of pacing themselves through the homework assignment also leads to better recall of the material. (Bartkovich & George, 1980, p.21)

For younger students, parents may need to supervise homework to ensure that it is completed each week. It is crucial, however, that the student be self-motivated and interested in participating in the DT→PI process, and is not doing it simply to please the parents. Mentor and parents should try to make the work interesting and stimulating. Youths should find it much more fun and more ego-enhancing than regular math classes in school.

The DT→PI process is facilitated if the mentor has an overall plan and is a skilled evaluator of the actual progress the mentee is making. Setting particular goals and outlining a linear progression of topics to be covered are part of developing a systematic plan for the student. The item profile chart completed in Step 4 is one device for organizing the plan of study.

It is not necessary for mentors to work with only one student at a given time. It is possible for skilled mentors to group as many as five students together for mentoring sessions. Students who are at about the same level in their understanding of mathematics, even if they are not the same age, may be grouped for mentoring sessions. For example, the Center for the Advancement of Academically Talented Youth (CTY) at Johns Hopkins University offers its "Young Students

Classes" based on the DT→PI model. Elementary students are placed in small groups of two to five according to their performance on the diagnostic tests (Moore & Wood, 1988). Students in these groups benefit from appropriate placement within the mathematics curriculum as well as from the opportunity to interact with their intellectual peers. Nevertheless, a one-on-one setting may be the preference of many students and mentors.

Although diagnostic testing is an important component of the DT→PI model and has been the focus of this chapter thus far, the purpose of this model is not merely to test the students, but rather to provide appropriate instruction based upon their needs. A much larger proportion of time is spent in instruction, learning, and practicing compared to the amount of time spent in testing. The point is to go beyond testing by using the diagnostic test results to plan instruction.

Although the DT→PI approach to education has been successful with many mathematically talented children, it is not the only approach available for talented students. It can be thought of as an option for parents, teachers, and students to consider. This mentoring procedure provides an alternative to wholesale grade- or subject-skipping and avoids inappropriately placing the young student who has advanced cognitive skills into a grade that may be higher, but where math is taught at a pace more suitable for average students.

USING THE DT→PI MODEL

Two programs that use the Diagnostic Testing → Prescriptive Instruction model with elementary students are the Julian C. Stanley Mentor Program (JCSMP) and the Investigation of Mathematically Advanced Elementary Students (IMAES). The goal of both JCSMP and IMAES is to identify mathematically talented students, give them challenging work, and encourage them to move ahead as their abilities and motivation allow. To help parents and teachers understand the DT→PI model as it is used in the IMAES and JCSMP programs, we have listed our responses to some concerns about the process as well as explained some of the details of developing the programs (see also Lupkowski, Assouline, & Vestal, 1992).

WHO IS ELIGIBLE FOR THE MENTOR-PACED PROGRAMS?

Students in the upper elementary grades who score at the 95th percentile or above on a mathematics section of a nationally standardized test (such as the *Iowa Tests of Basic Skills*) are good candidates to take an out-of-level aptitude test and to be considered for special programming. The lower level of the *Secondary School Admission Test* (SSAT), which was developed for students in grades 5-7, serves as the out-of-level aptitude test. Students take the Quantitative portion; those scoring at the 50th percentile or above when compared to students two years older on the out-of-level SSAT are recommended for further diagnostic testing of their specific level of achievement. Mentor-paced instruction is based upon the results of that assessment (Assouline and Lupkowski, 1992).

In addition to earning high scores on aptitude and achievement tests, students who participate in the mentor-paced programs should be mature and highly motivated. Participation in the program should be reserved for children who are eager to study more mathematics; the program demands a great deal of time and effort on their part. It is critical for children who participate in such a demanding program to want to be a part of it, not responding to parent's or teachers' desires.

A third consideration for placing students in the mentor program, in addition to student's aptitude and motivation, is the level of cooperation of the school. For the most part, schools have been eager to have their students participate in these special programs because they understand that JCSMP and IMAES serve as a supplement to what schools are able to provide. One of the most important ways in which that cooperation is assured is through communication. Letters from IMAES and JCSMP staff explaining the program and meetings with parents, mentors, and school personnel facilitate that communication (See Appendix A).

HOW IS THE BEGINNING INSTRUCTION LEVEL DETERMINED?

A student's knowledge of mathematics is assessed using a battery of achievement tests. The battery of out-of-level achievement tests includes the first two levels of the Basic Concepts and Computation tests of the *Sequential Tests of Educational Progress* (STEP), and tests from the *Cooperative Mathematics* series (beginning with Structure of the Number System and Arithmetic all the way through Calculus). Students scoring at a high level on these tests may participate in a continuum of activities based on their interests, needs, and abilities. Classroom teachers who are interested in applying this model might consider using chapter tests and final examinations provided by textbook publishers for diagnosing what a student has already mastered or has yet to learn.

It may seem as though a great deal of time is spent in administering standardized tests. However, only a small portion of the total time devoted to the mentor-paced program is allocated to standardized testing and most of it is in the beginning stages of the program. Once instruction begins, the time between standardized testing sessions is usually several months, although mentors continuously test children via chapter tests and teacher-made tests to be certain that they understand the material they are studying.

One of the children who participated in a mentor-paced program was "Matthew," a 10-year-old fifth grader who had been studying fifth grade mathematics. He qualified for participation in the program by scoring at the 92nd percentile on the Quantitative section of the SSAT when compared to 7th grade boys. Clearly, Matthew has exceptional abilities in mathematics. During the spring of his fifth grade year he took a series of achievement tests; on one test in which he was compared to 8th graders he scored at the 89th percentile on Basic Concepts and at the 85th percentile on Computation. Based on the results of diagnostic testing, Matthew was matched with a mentor who helped him fill in his few gaps in elementary school mathematics. Then he moved into the pre-algebra curriculum (using Merrill's *Pre-Algebra* text). By the time he entered sixth grade in the fall, Matthew had completed the pre-algebra textbook and was ready to begin Algebra I. A more complete description of Matthew's participation in the mentor-paced program is found in Chapter 6.

WHO ARE THE MENTORS?

We recommend that mentors be adults such as trained mathematics teachers, engineers, college professors, graduate students in mathematics, or mathematics education majors. In addition to a strong background in mathematics, the mentor needs to have the maturity to work with younger students and a good rapport with talented youth. Sometimes it is suggested that a high school student might serve as an effective mentor. However, our experience has been that high school students typically are busy with their own course work and extracurricular activities, and thus would have difficulty managing the responsibilities of planning for the mentoring sessions and meeting consistently on a weekly basis.

Typically the mentor-to-student ratio is 1:1. However, skilled mentors can work with small groups of students (2-5) who are at about the same level in their understanding of mathematics (Moore & Wood, 1988).

WHO COORDINATES THE MENTOR-PACED PROGRAMS?

The mentor program coordinator might be a university staff member, a gifted education teacher, or another member of the school staff who is familiar with standardized testing procedures and with mathematics curriculum. The coordinator's role is to identify students and assess their level of mathematical knowledge via aptitude and achievement tests and then match them with appropriate mentors. They regularly consult with mentors concerning students' progress and help mentors find appropriate materials to use with their students. Another important role is to oversee the administration of post-tests and keep administrators and classroom teachers informed of the students' progress.

HOW OFTEN DO MENTORS AND MENTEES MEET?

Typically, mentors and students meet once a week for two hours throughout the school year. During that time, they go over homework assigned during the previous meeting, clarify questions, and investigate new topics. One two-hour time period per week is preferable to several shorter time periods because it permits mentor and mentee to study topics in greater depth without interruption. We recognize that this schedule might not fit easily into the school day; it has been necessary to plan after-school and weekend sessions for many of the children in the JCSMP and IMAES programs. However, children often work on their mentor-assigned homework during the daily mathematics period while other children are doing their seat work. One school system that was extremely eager to have a student participate in the mentor-paced program allowed the mathematics mentor to go into the classroom twice a week for an hour to work with a student.

WHAT HAPPENS BETWEEN MEETINGS?

Although the student and mentor typically work together only once per week, students should be assigned homework to be completed daily prior to the next session. Students participating in the mentor-paced process should be self-motivated and self-disciplined, but parents still should be actively involved in supervising the completion of homework. Students should feel free to contact the mentor by phone if they have questions concerning mentor-assigned homework. Some mentors find that designating a particular day of the week for a phone call from the mentee is helpful.

WHO SHOULD ADMINISTER THE ACHIEVEMENT TESTS THAT ASSESS THE STUDENT'S PROGRESS?

It is preferable that the tests be administered by the mentor or in collaboration with school personnel to insure that the testing is conducted under standardized conditions. Achievement testing is necessary to insure that students receive proper credit for work done, since that work is not part of the regular curriculum.

All testing should include a written report of the results, a description of the child's behavior during the testing, and recommendations for prescriptive instruction based upon the results. The report should be shared with parents as well as appropriate school personnel (e.g., principal, counselor, and other teachers). More information about this type of report is found in Chapter 6.

WILL STUDENTS PARTICIPATING IN THE MENTOR-PACED PROGRAM HAVE LESS OPPORTUNITY TO INTERACT WITH OTHERS ABOUT THE MATERIAL BEING STUDIED?

Although students who participate in a mentor-paced program typically spend a great deal of time working alone, the mathematics they study is not learned in a vacuum. Because they are working in a one-on-one or small-group situation, student and mentor can toss around ideas about the material being studied. The mentor will be able to provide more accurate answers to questions than peers can; also, the mentor has the knowledge and ability to extend the student's questions and ask new questions of that student.

Another concern is that students will have less opportunity to interact with their peers and may even be "isolated" from them by participating in the mentor-paced process. Remember, however, that mathematics class is only one place in which students interact with each other. We strongly encourage talented students to participate in mathematics clubs and contests (e.g., MOES; see Appendix B), which give them a chance to interact with other students in a mathematics context. We encourage them to participate in unstructured play as well as more structured activities such as sports, music, etc. These varied activities will give them an abundance of opportunities to spend time

DT→PI

with their peers.

The time spent interacting with peers should be balanced by adequate time alone. It is not necessary for students to spend every spare moment interacting with peers. Allowing talented students to have plenty of independent thinking time is important; children need time to reflect, plan, and dream. This "alone time" allows talented students the opportunity to think about what they have learned and experienced. The time can be used to formulate new questions and to investigate them. Thus, what is sometimes perceived as isolation by adults actually has many positive aspects.

WHAT SHOULD STUDENTS WHO PARTICIPATE IN THE MENTOR-PACED PROGRAM DO WHILE THEIR CLASSMATES ARE IN THE REGULAR MATHEMATICS CLASS?

Before beginning a mentor-paced program, educators, parents, and administrators need to discuss the options available for a student while his or her classmates are in math class with their regular teacher. These options include: reading, working on the computer, going to the library, and working on mentor-assigned homework. Students in a mentor-paced program should be included for classroom games and other group activities. However, it is inappropriate to require students to sit through the regular mathematics class if they are already working with a mentor on more advanced mathematics. Not only would the talented student be bored in the regular class; he or she may also be disruptive.

If the student works on mentor-assigned homework while the other children are in the regular mathematics class, he or she will need a quiet setting with adult supervision. It is important to structure that situation so that the student does not feel that he or she is being ostracized or punished. Not only should parents, mentors, and school personnel discuss this, but also the students should be made fully aware of the reasons for the special arrangements.

MANY TEACHERS WORRY THAT REMOVING TALENTED STUDENTS FROM THEIR CLASSES WILL REMOVE AN IMPORTANT ROLE MODEL FOR AVERAGE STUDENTS.

This is a valid concern, especially in light of the cooperative learning movement that is gaining momentum in schools throughout the country. This concern is also mirrored in statements that question the validity of any type of ability grouping. Ability grouping and cooperative learning, as they interface with gifted education, were discussed in Chapter 3.

To be effective, role models must be somewhat close in ability to those who would benefit from exposure to the models (Schunk, 1987). Large differences in ability may promote arrogance on the part of the high-ability student. If teachers teach a mixed-ability group at the appropriate pace and depth for the extremely mathematically talented student(s), less able students experience unnecessary pressure from a set-up for failure. Mixed-ability grouping in mathematics, with extreme variation in aptitude, increases the management problems of teachers. However, teachers of students who participate in a mentor-paced program can concentrate on teaching the majority of their students and can be satisfied that their highly-able students' needs are being met.

HOW ARE ISSUES OF GRADES, CREDIT, AND PLACEMENT RESOLVED?

Before children begin participating in the mentor-paced program, parents should visit with school personnel to determine how children will receive grades and credit for the mathematics work that they will be doing. Each school resolves these issues in its own way. For example, some make special notes on children's report cards indicating what grade level mathematics students are studying. Mentors continually assess students' progress, so they are frequently asked to turn in grades to teachers. In other cases, mentors give chapter tests to teachers to administer, and that is how a grade for a particular marking period is determined. It is important to maintain careful records of students' progress so that, when the time comes to make a placement decision, students will be placed at the proper level in mathematics.

IT SOUNDS AS IF STUDENTS WHO PARTICIPATE IN THE MENTOR-PACED PROGRAM WILL BE ACCELERATED. WHAT HAPPENS THEN?

The mentor-paced program is based upon a process of diagnostic testing and prescriptive instruction that is designed to move students through the curriculum at a challenging pace. By definition, mathematically talented students will be accelerated in mathematics. Thus, the importance of planning before entering this process cannot be overemphasized. For example, elementary students who have participated in a mentor-paced program may be ready to take a high school mathematics class long before they are high school students. Special arrangements such as transportation from one school to another may be needed for these students. Older students may complete all of the available mathematics courses a year or more before graduating from high school. For these students, arrangements may be made with a local college or university so that the study of mathematics will not be interrupted. These and other issues are discussed in Chapter 8.

FOR HOW LONG DO STUDENTS PARTICIPATE IN THE MENTOR-PACED PROGRAM?

Because the mentor-paced programs are individualized, the length of time that students participate is very flexible. The mentor-paced program is usually used as a bridge to prepare students to take the next level of mathematics in their own schools. One student was almost ready to take an Algebra II class; she spent the last four weeks of the summer "cleaning up" her knowledge of Algebra I with a mentor in order to be ready to take Algebra II in her school. Other students have participated in the program for a semester or a full school year. One exceptional student who was studying geometry as a fifth grader will probably participate in a mentor-paced program for a number of years while he takes other subjects with his age-mates. If students who have completed the mentor program and are taking a regular class find at some future point that they are again well beyond their classmates in mathematics, they can re-enter the mentor-paced program. We strongly suggest that students re-enter a classroom setting (most likely with older students) as soon as practical.

IS THERE A COST TO THE PROGRAM?

Parents are generally expected to bear the financial burden of the mentoring when it occurs outside of school. For example, in the JCSMP, parents paid an hourly tutoring fee in addition to the costs of books and materials.

WHAT EFFORTS ARE MADE TO COOPERATE WITH SCHOOLS?

Most of the school personnel who have become aware of the mentor-paced programs have been extremely cooperative. In many cases, teachers have been the first to recognize talent and have encouraged parents to contact program administrators. In other cases, parents were the first to investigate the program. Parents are always encouraged to provide school personnel with information including reprints of pertinent articles and test reports and to discuss educational options for their child. Form 1 in Appendix A illustrates the letter sent to school personnel describing the JCSMP and inviting their support of the program for children who have qualified.

Programs such as IMAES and the JCSMP are conducive to a system of "dual mentoring" as described by Clasen and Hanson (1987). In dual mentoring, one mentor (the classroom teacher) attends to the developmental needs of the youngster and the other (the mathematics mentor) attends to the mathematics needs. Thus, the two mentors work in tandem to foster both the intellectual and the socio-emotional development of the youngster. The expertise of these two mentors is supplemented by university personnel who administer the program and train the mentors.

COULD THIS MODEL BE USED IN OTHER SUBJECT AREAS?

The mentor-paced approach is useful for subjects other than mathematics. For example, students have mastered the topics of high school biology, chemistry, and physics in mentor-paced programs (see Stanley & Stanley, 1986). Other subjects that lend themselves well to this approach include foreign languages, grammar, and writing skills. The same process of out-of-level diagnostic testing followed by prescriptive instruction can be applied. The creative mentor can assist a student in moving systematically through a course at a challenging depth and pace.

WHAT TYPES OF ACCOMMODATIONS MIGHT BE ESTABLISHED FOR STUDENTS WHO ARE TALENTED IN MATHEMATICS, BUT WOULD NOT BENEFIT FROM A MENTOR-PACED APPROACH?

The mentor-paced approach is appropriate for very talented students who are highly motivated to move ahead in mathematics. For those students who are well above average when compared to their age group but not extremely talented, other accommodations could be made. For example, teachers might group children by ability so that they could work together at an appropriate level and pace. This type of arrangement is frequently used in reading programs and lends itself well to the study of mathematics.

We recommend that parents and classroom teachers enrich the mathematics education of those students by using books and magazines, mathematics contests, computer programs, and other activities. Some of the materials available to schools provide excellent sources for mathematics enrichment. Additional suggestions for enriching the mathematics curriculum for talented students are included in Chapter 5.

CONCLUSION

In the DT→PI model, the study of mathematics is mentor-paced, so each student covers the material at an appropriate rate. Because students do not have to study topics they already know well, they have more time to study challenging topics with which they are not familiar, in greater depth than would be possible in the regular classroom. The model is self-correcting because students do not move on to the next topic before mastery is attained. It avoids the pitfall of the "gaps" that may occur with grade-skipping, while ensuring that the student is challenged by the pace at which material is presented and stimulated by the more advanced material.

Extremely talented students who participate in the DT→PI process are consistently presented with challenging material. Therefore, they avoid developing bad habits such as not writing out or checking their work when solving problems or doing complicated computations. Another advantage is that students learn that testing can be used for reasons other than evaluation, i.e., testing in the DT→PI model is used to determine appropriate topics and levels of instruction.

Before starting the DT→PI procedure, the student, parents, and mentors should work in concert with the classroom teacher. All of the people involved must consider the ramifications of participation in the DT→PI procedure. For example, important issues to consider are how the student earns school credit and advanced placement in mathematics for work completed with the mentor and how the mentoring experience will fit in with the school's curriculum and grading procedures.

The DT→PI model allows students who are exceptionally able in mathematics to progress at an appropriately accelerated rate through a mathematics curriculum. These few students will move more rapidly through the school's curriculum than is typical. This can be considered an accelerative model, or what David Elkind (1988) might call "tailoring." He says,

Promotion [in grade placement or subject matter] of intellectually gifted children is simply another way of attempting to match the curriculum to the child's abilities, not to accelerate those abilities. Accordingly, the promotion of intellectually gifted children in no way contradicts the accepted view of the limits of training on development, nor the negative effects of hurrying. Indeed, the positive effects of promoting intellectually gifted children provide additional evidence for the benefits of developmentally appropriate curricula. (p. 2)

Table 4.1

The DT→PI Model for Secondary Mathematics

In the DT→PI process as used with talented junior high students, the 7th or 8th grader begins by taking an *aptitude test*, the mathematical portion of the College Board's Scholastic Aptitude Test (SAT-M). A score of at least 500 on the SAT-M before age 13 is recommended to continue with this procedure. In the second step, *achievement pre-testing*, the student takes a standardized test of first-year algebra, such as the Cooperative Achievement Test, Algebra I (available from CTB/McGraw-Hill, 2500 Garden Road, Monterey, CA 93940). When testing is completed, the mentor scores the responses. If the percentile rank of the examinee is at least 50 on national eighth-grade norms, the mentor proceeds with Step 3, *readministering the missed items*. The examinee is given a list of items he or she marked incorrectly and is asked to try them again and show all work. The mentor examines the items missed a second time and asks: (a) did the examinee miss the items the same way both times and (b) what are the underlying principles of those items? These questions aid the mentor in diagnosing what parts of Algebra I the examinee does not know and in implementing the fourth step of the model, which is *prescribing an instructional plan* based upon points not yet well understood. The mentor does not require the student to work through an entire Algebra I book; rather, he or she asks the student to do suitable problems concerning the topics not yet well known (Stanley, 1978). The fifth and final step of the DT→PI model is a *post-test*. The objective is for the student to earn a nearly perfect score on a parallel (independent but comparable) form of the same test, thus indicating mastery of the material. The second through fifth steps are repeated for subsequent levels of the achievement test, such as Algebra II and Algebra III.

DT→PI

MATHEMATICS CURRICULUM AND MATERIALS

OVERVIEW

The goal for developing curricula for mathematically talented elementary students is to provide a challenging, content-filled program that allows them to move steadily through mathematics. The content should include many opportunities to reason, relate ideas, formulate and solve problems, become competent with computers and calculators, and learn when and how to estimate, while establishing a firm foundation of mathematics concepts, principles, and rules. In this chapter, a rationale for developing a differentiated curriculum and resources for implementing such a curriculum are presented. Additional information about materials is listed in Appendix D.

Parents, mentors, and teachers typically use one of four models when attempting to differentiate for mathematically gifted students (Pratscher, Jones, & Lamb, 1982). These include: (1) Enrichment Model, (2) Breadth/Depth Model, (3) Topics Model, and (4) Acceleration Model (as advocated by the Study of Mathematically Precocious Youth).

THE ENRICHMENT MODEL

The Enrichment Model is helpful for students who have above average abilities in mathematics, but are not exceptionally talented. It is often used in heterogenous classrooms. One favorite focus of enrichment programs is computer science; computers might be used to solve problems such as finding all of the prime numbers between 0 and 200 or figuring out how many ways a person could make change for a dollar.

Curriculum

THE BREADTH/DEPTH MODEL

The Breadth/Depth Model is also helpful for use in a heterogeneous classroom. The depth into which a particular topic is pursued varies from student to student. "For example, in an Algebra I class, all students would be expected to use the quadratic formula, but the gifted students would also be expected to derive it. In a junior high school math class, all students would be expected to do computation. The gifted students, however, could spend the time doing computation in bases other than base ten" (Pratscher, Jones, & Lamb, 1982, p. 367). This model is probably most useful for those students who are talented in mathematics, but not extremely so (e.g., those who score in the top five percent on a grade level standardized test but below the 50th percentile on tests developed for students two or three grade levels above; see pages 33-34, Chapter 4). The breadth/depth model avoids the problem of gifted students being given "more of the same" busy work. Instead, the students are given a more in-depth experience at each level of instruction. Another advantage of this model is that it is relatively simple for the classroom teacher to deliver and does not require the development of a totally separate program for the gifted. Instead, the teacher can match in-depth activities with each level of the existing curriculum.

TOPICS MODEL

In typical elementary school programs, the focus is on learning computation skills. For talented students who are grouped together, instructors may focus instead on teaching topics such as probability, number theory, statistics, and geometry. Computation skills can then be taught as they relate to the topic currently under study, and students will immediately understand the relevance of a given computation skill. For example, during a unit on statistics, students might learn computation of decimals. "This model gives the teacher the opportunity to use a diagnostic/prescriptive approach in teaching computational skills, while devoting most of the class time to the more interesting and motivating aspects of mathematics. The topics model is most effective in the homogeneous situation, since it involves major curricular adjustments" (Pratscher, Jones, & Lamb, 1982, p. 369).

THE ACCELERATION MODEL

The Acceleration Model (also known as the Diagnostic Testing → Prescriptive Instruction Model) was described in detail in the previous chapter.

Curriculum

WHAT HAPPENS IN A TYPICAL CLASSROOM?

In an effort to document what is taught in elementary mathematics classrooms, Porter (1989) examined teacher logs and interviewed elementary teachers. From his study, three features of elementary school mathematics instruction emerged:

(1) Elementary mathematics instruction emphasizes skill development; basic concepts and problem solving receive only cursory attention. This emphasis is mirrored in the textbooks.
(2) A large percentage of topics receive only brief coverage. Seventy percent of the topics receive less than a total of 30 minutes throughout the school year.
(3) There is a remarkable slowness with which content changes from grade to grade.

Our experiences with children throughout the country suggest that these findings are generalizable and are not specific to the locale in which the research was conducted. The National Council of Teachers of Mathematics (1989) also notes that there is little change in textbooks from grades 5 through 8. Flanders (1987), in his examination of three commonly used textbook series, found that only 30 to 49 percent of the content in 4th through 8th grade textbooks was new. "It is even more disconcerting to realize that the very chapters that contain the most new material, such as probability, statistics, geometry, and pre-algebra, are covered in the last half of the books-- the sections most often skipped by teachers for lack of time. The result is an ineffective curriculum that rehashes material students have already seen" (NCTM, 1989, p. 66).

Because children need to know number facts and to be able to manipulate numbers correctly, the typical elementary mathematics curriculum has evolved to the current practice of emphasizing skill development of the basic number facts. This practice is reflected in the content of textbooks that are marketed and used throughout the United States. However, even elementary school mathematics should consist of much more than rote learning of computational skills. In addition to learning number facts and becoming facile with computation, the National Council of Teachers of Mathematics recommends that all elementary students should be exposed to areas such as problem solving, applying mathematics in every day situations,

Curriculum

estimation and approximation, geometry, measurement, interpreting and constructing charts and graphs, using mathematics to predict, and computer programming (NCTM, 1980). For mathematically talented elementary students, Wheatley (1983) recommends that the following areas should be studied: problem solving, geometry and measurement, facts and computation, arithmetic and algebraic concepts, computer programming, numeration, probability and statistics, estimation, and spatial visualization.

We recommend that students should receive more than just exposure to these topics. Merely exposing them to such concepts is likely to convey two unintended messages: (1) that concepts and applications are not as important as speed and accuracy and (2) knowing just a little bit about a variety of topics is more important than a thorough understanding of key concepts (Porter, 1989).

NCTM STANDARDS

In 1989, the National Council of Teachers of Mathematics (NCTM) published its *Curriculum and Evaluation Standards for School Mathematics* as one means to help improve the quality of mathematics education in the United States.

> We believe that all students should have an opportunity to learn the important ideas of mathematics expressed in these standards. On the one hand, prior to grade 9, we have refrained from specifying alternative instructional patterns that would be consistent with our vision. On the other hand, for grades 9-12, the standards have been prepared in light of a core program for all students, with explicit differentiation in terms of depth and breadth of treatment and the nature of applications for college-bound students. (NCTM, 1989, p. 9)

Thus, the elementary school mathematics curriculum that the NCTM advocates does not make specific recommendations for differentiating for talented youth. Instead, all children are expected to study the same basic material and it is up to individual schools and teachers to develop appropriately challenging programs for their mathematically talented youngsters. They do recommend in an earlier publication that ". . . programs for the gifted should be based on a sequential program of enrichment through more ingenious problem solving opportunities rather than through acceleration alone" (NCTM,

1980, p. 23).

According to the NCTM (1989), the curriculum for children in kindergarten through 4th grade should emphasize learning concepts, doing mathematics (instead of memorizing facts), and developing thinking and reasoning abilities, as well as emphasizing applications, including a broad range of content, and making use of calculators and computers.

In grades 5 through 8, the curriculum would "expand students' knowledge of numbers, computation, estimation, measurement, geometry, statistics, probability, patterns and functions, and the fundamental concepts of algebra" (NCTM, 1989, pp. 65-66). Increased attention should be given to pursuing open-ended problems, discussing mathematical ideas, reasoning, connecting mathematics to other subjects and to the outside world, and creating algorithms and procedures. Decreased attention should be paid to memorizing rules and algorithms, practicing tedious paper and pencil computations, memorizing formulae, and learning topics in isolation. The focus now should be on real-world problems, such as those that "allow students to experience problems with 'messy' numbers or too much or not enough information or that have multiple solutions, each with different consequences, will better prepare them to solve problems they are likely to encounter in their daily lives" (NCTM, 1989, p. 78).

CONCEPTS VS. COMPUTATION

Mathematically talented students are often penalized by not being allowed to do more interesting, advanced work because their careless errors in computation (often a reflection of being bored with the nature of the task) are taken as proof that they are not ready for more advanced work.

"One of the quickest ways to turn very able children off to the excitement of learning is to bore them with mundane repetition of skills they have long ago mastered" (Wolfle, 1987). It is especially important not to require mathematically talented students to spend an excessive amount of time memorizing number facts, because they are more likely to learn rote material quickly and need the challenge of studying material that is more conceptual in nature. For example, Wheatley (1983) suggests that it is not necessary to wait to introduce problem solving and advanced topics until the computations have been mastered. Rather, computational skills should be taught concurrently with concepts. In this way, children will understand why the computational skills are needed and will learn them more efficiently. We have found a trend that supports Wheatley's recommendation; the mathematically talented elementary students with whom we work tend to be more advanced in their understanding of concepts compared to their ability with computational skills. For example, when mathematically talented students are tested with Sequential Tests of Educational Progress, their Basic Concepts scores are generally higher than their Computation scores. On standardized tests such as the ITBS, the same pattern is often seen. Mathematically talented students usually have higher scores on basic concepts and problem solving subtests than on computation subtests.

The National Council of Teachers of Mathematics (1989) also opposes the notion that students should be prohibited from studying a broader curriculum until they have mastered the basic computational skills. They support shifting the focus to a broader curriculum for the following reasons:

(1) Computational proficiency is not the only "basic skill" today. Because of the increase in technology, topics such as geometry, probability, statistics, and algebra are more accessible and more important for students.
(2) The methods used in the past for teaching students basic computational skills have not been as effective as one would hope. Why should students be successful now if the same methods are still used?
(3) By studying mathematics topics and applying what they have learned, students will understand the need for learning arithmetic concepts and skills such as adding and multiplying fractions.

TEN AREAS OF MATHEMATICS FOR TALENTED ELEMENTARY STUDENTS

Wheatley (1983) proposed that ten areas of mathematics should be studied in varying degrees in elementary school. His strands are listed in Table 5.1, and will be explained in detail in this chapter. Not only should talented students study all of these areas, but also the curriculum for them can be enriched in these areas. Once students have mastered topics in a given area, they are ready to move on to the next level.

Table 5.1
Ten Mathematics Strands (from Wheatley, 1983)

Strand	Percentage of time to be devoted to each
Problem Solving	20
Geometry and Measurement	15
Facts and Computation	15
Arithmetic and Algebra Concepts	12
Computer Programming	10
Estimation and Mental Arithmetic	6
Numeration	6
Probability and Statistics	6
Spatial Visualization	5
Applications	5
Total	100

PROBLEM SOLVING

The ability to solve problems does not develop automatically as children master computational skills; they need to be taught problem solving skills (Lenchner, 1983). Solving problems is different from doing an exercise in mathematics. Exercises are used to practice a known procedure; the means to arriving at the solution of an exercise is evident. In contrast to an exercise, the solution or the means to solving a problem is not always apparent, and generally requires some degree of originality or creativity on the part of the problem solver (Lenchner, 1983). Thus, problem solving is a skill requiring a more sophisticated level of cognitive functioning as compared to the level required for basic computation.

Teachers might begin to teach problem solving by teaching problem-solving heuristics. Appropriate heuristics for elementary students include: drawing diagrams, making lists, finding patterns and comparing the problem to problems encountered previously, working backwards from the solution, and testing and verifying (Lenchner, 1983). Some problem solving activities provide an excellent means for students to work cooperatively. When children solve problems in groups, they can learn from interacting and discussing the problem with others. By describing how they solved the problem, students learn about the thinking processes they have gone through.

Curriculum

There are many books on problem solving that are readily available to students, parents, and teachers. Three such books are *Creative Problem Solving in School Mathematics* by Lenchner (Boston: Houghton Mifflin, 1983), *Teaching Problem Solving: What, Why and How* by Charles and Lester (Palo Alto, CA: Dale Seymour Press, 1982), and *Problem Solving: A Handbook for Teachers* by Krulik and Rudnick (Boston: Allyn & Bacon, 1987, 2nd edition). Helpful problem solving books include sample problems, activities, games, and methods for teaching problem solving skills.

In order to enrich the elementary mathematics curriculum for talented students, many schools have turned to the abundant materials available on problem solving and are using problem solving activities as their "gifted program." Although problem solving activities are challenging and fun for talented children, they do not provide all of the mental stimulation children need; problem solving is not the one answer to the gifted student's needs in mathematics. In fact, no single area can provide the challenge that talented students need; they need a complete, balanced program in mathematics that moves them along in a systematic fashion.

GEOMETRY AND MEASUREMENT

Instead of studying formal theorems or proofs, elementary students would benefit from a study of geometry that emphasizes concepts and principles instead of the focus on definition of terms and shapes that previously characterized elementary students' study of geometry. This will provide the foundation for the formal geometry courses that talented students will take at a later point. One text that may be appropriate for youngsters who are not ready for a formal course in high school geometry is *Experiencing Geometry* by Bruni (1977, Belmont, CA: Wadsworth Publishing Company). Bruni's book is an informal, intuitive introduction to geometry. It contains numerous concrete models and illustrations and encourages readers to investigate fundamental geometric concepts through observation and experimentation.

In addition to using textbooks to study geometry, it is especially important that elementary children use manipulatives. The use of manipulatives makes an abstract topic more concrete and understandable. For example, students can calculate the area of a rectangle using unifix cubes. "Surface area and volume of prisms, pyramids, and other polyhedra provide a rich setting for establishing relationships and building intuitions" (Wheatley, 1988, p. 257).

Students should also learn how to use a protractor and begin graphing functions. Wheatley (1988) recommends avoiding rote applications of formulae; instead, work with formulae should be approached meaningfully.

FACTS AND COMPUTATION

It is critical for elementary students to study and master the basic facts of arithmetic because further study in mathematics requires accurate calculations. Although calculators are readily available, children still need to know the basic addition facts, multiplication tables, etc., because they need to be able to estimate and verify their answers.

Although mastering facts and gaining facility with computation are critical for elementary students, it is not necessary to spend an excessive amount of time on these areas. Unfortunately,

> a survey of current texts will show that more than 50% of the school year is spent on computation and memorization of facts. In practice, most teachers spend closer to 70% of the year on these topics. A key to providing for the gifted and talented is limiting the time on computation. Gifted children not only learn faster but they retain what they learn. . . . The time for facts and computations can be reduced providing time for other topics. A diagnostic-prescriptive instructional strategy is particularly appropriate for this strand (Wheatley, 1983, p. 79).

The Diagnostic Testing → Prescriptive Instruction approach described in Chapter 4 is one way to determine exactly what level students have achieved in terms of knowledge of basic facts. Administering mastery tests will determine the level of mastery children have achieved. Then, students can move on to other areas and not be required to repeat or review material that they have already learned well. By using the diagnostic/prescriptive approach, more time is available for activities in the nine other areas of mathematics. Although this approach may not be appropriate for all talented students, an adaptation of the model can be used to assess whether or not students have mastered the basic computational facts. Once mastered, there is very little need for repeated practice.

Curriculum

Some talented children become adept at manipulating parents and teachers and convince them that they don't need to do any drill or repeated practice. Although drill can be relatively boring, it is needed occasionally. In one case, a talented girl complained of headaches whenever she was asked to practice the multiplication tables or if she didn't know the answer to a multiplication problem. By administering mastery tests, it was determined that she did not have a good knowledge of basic multiplication facts and required drill in these areas. However, she was not held back in other areas because of her lack of knowledge of those facts. Instead, she was given plenty of opportunities to practice computational skills while still working with more advanced concepts.

ARITHMETIC AND ALGEBRAIC CONCEPTS

Prime, composite, factor, divisibility, ratio, and proportion are some of the important arithmetic concepts that elementary students need to learn. Algebraic concepts that should be covered in elementary school include the concept of variable, signed numbers, equations of curves in a plane, and intuitive solutions of number sentences (Wheatley, 1983). To understand these concepts, manipulatives may be especially important and helpful to students. For example, students could use square blocks and triangular blocks as symbols to "write" and solve equations and problems such as: One square plus one square equals one triangle. How many squares are needed to make two triangles? Wheatley (1988) recommends the book, *Routes to Algebra,* by J. Mason and D. Pimm (1985, Milton Keynes, England: Open University Educational Enterprises) as a text for an algebra course based on understanding and meaning rather than special methods. He asserts, "Effort should be made to approach each topic as problem solving so that students can construct meaning for themselves" (p. 257).

COMPUTER PROGRAMMING

We are constantly reminded that we live in the "computer age" and realize the importance of helping children to be comfortable with computers. Knowledge of computers will be essential when today's youth are adults. Because children need to become computer literate, they must have opportunities to use computers. Not only is computer programming valuable for its own sake, but also it has the added benefit of helping children to practice logic. Writing and debugging programs require reasoning and discipline (Wheatley, 1983). Popular computer languages for children include LOGO and BASIC. LOGO can be used for developing logic skills, solving geometry problems, and exploring concepts. BASIC is a computer language with many uses; students who study and experiment with this language will be able to transfer their knowledge when learning more advanced languages in later years. Wheatley (1988) recommends spending a limited amount of time teaching a programming language; instead, he recommends that students be presented with problems to solve on the computer after learning only a few programming statements.

ESTIMATION AND MENTAL ARITHMETIC

Although calculators and computers are used more and more frequently to perform calculations, it is still necessary for children to learn to estimate their answers. They can evaluate the results obtained from calculators and computers and develop sensitivity to the "reasonableness" of an answer (Wheatley, 1983). Wheatley recommends the book, *Guide to Using Estimation Skills and Strategies* by Reys and Reys (Palo Alto, CA: Dale Seymour, 1983) for practicing estimation.

The Center for the Advancement of Academically Talented Youth (CTY), which coordinates the Young Students Program for mathematically talented elementary school students, recommends that children develop the habit of always checking out a problem before beginning any computation to determine approximately what their answers will be. After working out the answer, students should always ask themselves if it makes sense (Mills & Wood, 1988). The 1986 National Council of Teachers of Mathematics Yearbook, *Estimation and Mental Arithmetic*, contains helpful information on integrating estimation into the mathematics curriculum.

NUMERATION

An interesting way to enrich the study of numeration and number systems is to study the history of mathematics. For example, students can study the number systems used by ancient cultures and can even make up their own number systems. By studying other numeration systems, students will better understand the base ten system (Wheatley, 1983).

James M. Henle's book, *Numerous Numerals* (Reston, VA: National Council of Teachers of Mathematics, 1975), includes exercises on numeration systems such as "fracimals," "zerones," and "negaheximals" and problems on continued fractions and bases. Although the material was written for seventh graders, younger

Curriculum

talented students should be able to do the problems.

PROBABILITY AND STATISTICS

Rather than always thinking in absolute terms, talented elementary students can begin considering the chance of an event occurring by studying probability and statistics. Students can become involved in studying probability and statistics by gathering data and making tables and graphs. For example, talented preschoolers enjoy collecting data for surveys such as "What is your favorite kind of ice cream?," tabulating the responses, and making graphs of the results. Older children enjoy more complex problems, such as surveying a sample of adults and predicting the results of a local election. In addition, students might study the probability involved in card or dice games or sports (Pratscher, Jones, & Lamb, 1982). Children will learn that concepts of probability can be applied to chemistry, physics, and economics; these subjects are likely to be studied by gifted students, so an early introduction to these topics is helpful (Wheatley, 1988).

Elementary students will benefit from discussions of theory and applications of statistical techniques in business, medicine, culture, recreation, etc. They can learn about misleading statistics, sampling techniques, and concepts such as mean, median, and mode. By studying the patterns that emerge when they flip coins, toss dice, or spin spinners, students can "see" probability in real life.

The article by Jim Swift, "Challenges for Enriching the Curriculum: Statistics and Probability" (*Mathematics Teacher*, 1983, pp. 268-269) offers ideas about including statistics and probability in the regular school curriculum and lists useful resources.

SPATIAL VISUALIZATION

The ability to visualize spatial transformations such as mentally turning a cube is important for doing mathematics, science, and engineering. In order to practice in this area, students might take advantage of the excellent computer games that are available commercially. Many of these computer games are listed in this chapter in the section, "Computer Games and Programs." Before students use computers, they might manipulate real objects in order to develop an understanding of spatial relationships; physically manipulating objects will help them to be better prepared to manipulate them mentally. Making three-dimensional models using tangrams, tessellations, and graphing help children learn to think visually (Wheatley, 1988).

APPLICATIONS

Applying mathematical concepts to real-life situations gives meaning to the ideas studied (Wheatley, 1983). It is helpful if students can do problems that have meaning to them. For example, one boy who was very interested in baseball was able to get a great deal of practice in computation by calculating statistics on his favorite baseball players.

In using mathematics in everyday situations, children might look at problems in unusual ways or they might try to solve a problem for which they haven't been taught an algorithm. We strongly encourage parents and educators to permit as well as encourage children to develop their own algorithms. By discovering them on their own, children learn more and retain what they have learned. However, we stress the importance of teaching children correct notation. For example, one clever student "discovered" matrices, and developed a way to write down what he was doing with them. Although he manipulated them correctly, he used non-standard notation. Unfortunately, he was greatly confused when he tried to understand matrix problems that others had done, because he could not understand their notation.

The National Council of Teachers of Mathematics (1989) strongly supports the idea that all elementary school students should study mathematics that is linked to every day situations and applications. The use of manipulatives such as unifix cubes, pattern blocks, dice, and cuisinaire rods may be especially helpful when attempting to apply the mathematics learned to "real life" situations. More information about manipulatives is provided in the "Materials" section of this chapter.

OTHER TOPICS

In his 1988 chapter, Wheatley also explored rational numbers, ratio, proportion, and percent as topics that should be covered in the curriculum for mathematically talented elementary school students. By using physical representations such as those found in Creative Publications' software, *The Pattern Factory*, primary children can work on an intuitive and conceptual understanding of rational numbers, for example. Wheatley notes that proportional thought requires formal operational thought (in the Piagetian sense), so it should not be introduced prematurely. Much work in the sciences uses proportional thought as its basis, so gifted students who are interested in the sciences will find this topic especially useful.

MATERIALS

TEXTBOOKS

Most of the textbooks presently available in elementary schools are based upon the spiral curriculum. That is, they expose students to a topic and then return to it again and again, each time at a slightly more advanced level. This format is not the optimal one for gifted students, who do not require constant review and repetition in order to understand a concept.

We have had difficulty in finding textbooks to recommend that are not based on the spiral curriculum. In addition, ". . . present textbooks do not develop higher level reasoning but over-emphasize computational rules. . . . The current texts are built on the assumption that topics will be repeated from year to year. Typically, a sixth-grade text will contain extensive review and reteaching of each topic. In fact, little prior knowledge is assumed. This approach is particularly devastating for the gifted. They are forced to study material they already understand" (Wheatley, 1983, p. 79). Flanders (1987) examined three textbook series and found the average number of pages offering new material each year was quite low; for grade 4, only 45 percent of the material was new, while for grades 5 and 6 the percentages were 50 and 38, respectively.

We recommend that teachers choose texts that stress concepts, not computation. "Teaching arithmetic algorithmically to gifted students is dangerous. It conditions them to attend to procedures and to ignore the logical structure of mathematics" (Howley, Howley, & Pendarvis, 1986, p. 216). We also recommend that efficient pretesting of the early chapters of the book, which are usually review chapters, take place immediately so that students can move on to more challenging topics found in the later chapters.

Bartkovich and George (1980) suggest that mentors and teachers select textbooks that are concise, contain clear explanations so that students can learn many concepts without assistance, provide examples of completely solved problems, and contain large numbers of practice problems that could be assigned for homework. The answers to about half of those practice problems should be provided in the text so students can check their own work as well as have assignments to hand in. Finally, chapter tests for diagnosis and evaluation are essential.

Those who teach mathematically talented elementary students should selectively use portions of available textbooks. Since most current textbooks use the spiral approach and talented students do not usually require repeated drill, we caution against going through a textbook page by page. Rather, a teacher or mentor might have the student begin a chapter by taking the chapter test as a pre-test. If the student performs well on the chapter test, he or she does not need to spend a great deal of time on that chapter. Instead, the student should study those topics not yet well understood and then move on to the next chapter or topic.

> The assignment of a great deal of written work to a child who is under ten years old is not *necessary* in an efficient instructional sequence. For example, suppose an eight-year-old gifted child is working in a fifth grade math book at a rapid pace. There is no reason for the child to complete every exercise in the book. Exercises can be sampled. Suppose fifty problems are included in the book to allow students to practice the skill. Depending on the teacher's understanding of the child's abilities, five, ten, or fifteen problems can be assigned to the gifted child who is progressing rapidly through the text. This sample of the problems should, if the teacher's judgment is accurate, provide sufficient practice and permit the teacher to check the child's grasp of the principle involved. The assignment of all fifty problems not only would prove burdensome to the child, but also would make high-level learning the hostage of low-level learning. (Howley, Howley, & Pendarvis, 1986, p. 211)

Two groups who work with mathematically talented elementary students, the Center for the Advancement of Academically Talented Youth (CTY) at Johns Hopkins University and the Center for Academic Precocity at Arizona State University, use the first eight chapters of the Keedy and Bittinger book, *Basic Mathematics* (Menlo Park, CA: Addison-Wesley, 1987, 5th edition) and follow that with a pre-algebra text such as *Pre-Algebra* by Price, Rath, and Leschensky (Columbus, OH: Merrill, 1986). CTY distributes materials that explain the curriculum in greater detail. Write to CTY, Johns Hopkins University, Baltimore, MD 21218 for more information.

Mentors in the Julian C. Stanley Mentor Program (JCSMP), developed by the Study of Mathematically Precocious Youth at the

Curriculum

University of North Texas) have discovered that many of today's elementary school mathematics textbooks have eliminated the topic of sets. We strongly encourage the reintroduction of sets into the elementary school mathematics curriculum, as this knowledge is essential as students continue their study of mathematics. Mentors in the JCSMP have used chapters from textbooks published in the 1960's and 1970's as resources for this topic. They have also developed their own workbooks on sets, bases, and measurement.

Another book that the mentors in the JCSMP have found useful is *Basic Mathematics: Preparing for Algebra* by Brenda Wood, Linda L. Exley, and Vincent K. Smith (1990, Englewood Cliffs, NJ: Prentice Hall). The book was written as a review of elementary school mathematics for students who are not quite ready for algebra; mentors can easily select sections of the book that provide explanations and practice problems for talented young students who have not yet learned or need more practice on a particular topic, such as fractions, percents, and rational numbers.

Gateways to Algebra and Geometry, An Integrated Approach by John Benson, Sarah Dodge, Walter Didge, Charles Hamberg, George Milauskas, and Richard Rukin (1992, Evanston, IL: McDougal, Littell, & Company) differs from traditional pre-algebra textbooks because it is an integrated program. Each problem set includes a variety of problems, and each chapter includes an excellent feature entitled *Puzzles and Challenges.*

The chapter entitled "Activities to Stimulate Numerical Thinking" in *Young Children Reinvent Arithmetic* (Kamii, 1985) may prove useful in generating activities for elementary aged children. Educators may also benefit from Part III: Activities in *Young Children Continue to Reinvent Arithmetic* (Kamii, 1989).

ENRICHMENT BOOKS AND RECREATIONAL MATHEMATICS BOOKS

A great number of mathematics enrichment books are available. Teachers might use them as resources for practice problems or as a source of ideas to use in the classroom. We recommend seeking books that are written in a clear, comprehensible manner so that individual students might work through them on their own. A general rule of thumb is to find books that were written for students a few years older than the talented students who will be using them. These books should offer a challenge to bright students. Enrichment books may be selected in a number of ways. They might be chosen on the

Curriculum

basis of topics or themes. Other books might add a historical perspective to students' study of mathematics. Contest problem books also provide excellent problems for practice. Numerous mathematics enrichment materials are listed in Wheatley (1988). Mathematics enrichment books include:

Algebra Can Be Fun by Perelman (translated from the Russian by George Yankovsky, 1979, Moscow: Mir Publishers). This is not a beginning algebra text or a manual; rather, "it is a book for free reading. It is designed for a reader with some knowledge of algebra. . . . It is meant to develop in the reader a taste for algebra and problem solving, and also excite him [or her] to dip into algebra textbooks and fill in the blanks in his [or her] knowledge" (p. 5).

Mathematical Quickies by Charles W. Trigg (1985, New York: Dover Publications). The book is written for "good high school students." The problems are not organized in order of difficulty.

Motivated Mathematics by Evyata & Rosenbloom (1981, New York: Cambridge University Press). This book is a resource for mentors on applied mathematics. Topics include polynomials, logarithmic functions, exponential functions, trigonometric functions, limits, probability and statistics, analytical geometry and calculus, and linear algebra. Some sections of the book were designed for students in grades 5-8; later sections contain more advanced material.

Calculus By and For Young People by Don Cohen (1988, Donald Cohen, 809 Stratford Dr., Champaign, IL 61821) begins "I was teaching a class of teachers at Webster College in the 60's; one day, Judy Silver, a first-grade teacher, figured out the relationship between the derivative and the integral. I had tears come to my eyes, I was so excited. . . . I want to share this book with young people. . . , so they can get an early start thinking about these ideas, and not have to memorize a lot of formulas and notations without much understanding, as I did in my mathematics courses." (p. iii).

The Executive Director of MOES, Dr. Lenchner, wrote two books of problems that are excellent resources for mathematically talented students: *Creative Problem Solving in School Mathematics* (Houghton Mifflin, Boston, 1983) and *Mathematical Olympiad Contest Problems for Children* (Glenwood Publications, Oceanside, NY, 1990). Problem sets from MOES Competitions from 1989 and 1991 are found in Appendix B. Another book of mathematics contest problems is *The 1st Elementary Grades Math League Problem Book* by Steven R. Conrad and Daniel Fleger (1989; Math League Press, P. O. Box 720, Tenafly, NJ 07670).

Curriculum

Many other enrichment books are available. Appendix D lists publishers that market such books.

COMPUTER GAMES AND PROGRAMS

Computers cannot take the place of good teaching; however, the proper use of software can help students develop a better, richer understanding of mathematics. For example, using graph paper and pencil is one way to study the concept of slope. However, children who can actually manipulate the rise and run of a line on a computer screen can see directly how they affect the slope. This leads to a stronger understanding of the concept.

Computer programs and games are useful for reinforcing basic skills and practicing different strategies. The same program can be used by students of varying skill levels. For example, a simple game might be used by young students to generate data that will be used by more advanced students for more complex problems.

Because students need plenty of opportunities to practice with the computer and because they enjoy playing computer games, they will probably want to play the games outside of school hours. Students might be able to borrow computer programs from their schools or libraries to use on their home computers. We encourage students to take the programs home because they enjoy them and the extra practice time allows their skills to become even more developed.

For parents and school personnel who are preparing to purchase computer software, we recommend that they request both home and school catalogues from companies that produce computer software. In some cases, the programs that are marketed for home use are flashier, but actually less useful than the programs marketed for the schools. Computer programs marketed for the schools may be more flexible. In other cases, home and school software are identical, but the programs marketed for the home are less expensive because they do not include a backup copy or teachers' manual.

Some programs are not necessarily marketed as "mathematics software," but have a mathematics element. For example, *Where in the World is Carmen SanDiego* (published by Broderbund and available in many computer software stores) and *Oregon Trail* are two games that have mathematics applications. In *Oregon Trail*, players make choices and use logic and decision-making techniques as they travel across the U.S. *Where in the World is Carmen SanDiego* tests players' knowledge of geography and requires them to use a variety of reference materials as they play.

Curriculum

The programs listed below are available from Sunburst Communications (39 Washington Ave., Pleasantville, NY 10570). Grade levels indicated are those listed in the Sunburst catalog. Younger students who are talented in mathematics may find these programs appropriately challenging. For example, although *Gnee or Not Gnee* was designed for children in 3rd through 6th grade, it can be used successfully with bright 3- to 4-year-olds. The list of computer programs here is not meant to be exhaustive. In Appendix D, we list names and addresses of other companies that produce software suitable for this group.

Building Perspective (grade 4 to adult). Players work with 3 x 3, 4 x 4, or 5 x 5 areas depicting buildings of different heights, viewed from the side. Players can use only this view but must visualize what the view from the top should look like.

The Factory (grade 4 to adult). Players create geometric "products" on a simulated assembly line. It requires inductive thinking, visual discrimination, spatial perception, understanding sequence, logic, and efficiency.

The Super Factory (grade 6 to adult). This game requires problem solving and provides practice with spatial geometry. Players research and design products. They rotate a cube with different faces and attempt to duplicate that cube.

Gnee or Not Gnee (grades 3-6) provides practice with visual discrimination and rule formation.

The Geometric preSupposer (grade 5 to adult) provides a foundation for geometry. Students measure lengths, angles, areas, and circumference, and experiment with mathematical combinations of these measures.

The Geometric Supposers (grades 8-12) lets players explore the properties of geometric figures and learn geometry experimentally. For younger students, it is best to use with a good mentor.

How the West Was One + Three X Four (grade 4 to adult). Players predict a final outcome. It offers practice with order of operations, parentheses, addition, subtraction, multiplication, and division and promotes logical thinking.

The King's Rule (grade 4 to adult). Players must discover numerical rules to work through a king's castle. They must generate and test hypotheses.

The Royal Rules (grade 6 to adult) is more challenging than *The King's Rule*. Players create their own rules and challenge others to solve them. This game provides practice in hypothesis testing and

Curriculum

helps students learn that the answer "no" can be useful (contrapositive).

The Pond (grade 2 to adult) allows players to recognize patterns, generalize from raw data, and think logically.

Puzzle Tanks (grade 3 to adult). Players have two tanks of varying capacities and a storage tank. They arrive at the target amount by filling and emptying tanks and transferring material from tank to tank.

Safari Search (grade 3 to adult). Twelve search activities on a 5 x 5 grid encourage inference. Players construct hypotheses, collect data, and make complex judgments. Players learn the concept of locus and are given a foundation for geometry.

Sir Isaac Newton's Games (grade 4 to adult). Formulae governing motion are recreated in simple games. Players can comprehend the basic concepts of motion as developed by Newton. They learn the effects of varying environments on motion, vector addition, and gravitational fields.

The Trading Post (grade 2 to adult). Two players analyze their starting positions and plan strategies to reach a specific goal. Players make trades to balance "equations" while learning pre-algebra skills, visual discrimination, and critical thinking skills.

The Voyage of the Mimi combines videos, computer software, and print materials. Students learn about whales, navigation, maps, and computers in this integrated program. It uses a LOGO syntax, and helps to develop readiness skills for computer programming.

The following programs are available from the Learning Company (6493 Kaiser Dr., Fremont, CA 94555):

Bumble Games (grades K-3) teaches the principles of basic mathematics and geometry. Players use number lines and number pairs and learn graphing skills.

Gertrude's Secrets (grades K-3). Players use basic thinking and problem solving skills while using pattern blocks to classify, group, and sequence.

Gertrude's Puzzles (grades 4-7) encourages players to think analytically. They order and sort pieces by shape and color and learn to recognize patterns and relationships. Players test and discover multiple solutions.

Rocky's Boots (grades 4-8) requires logical thinking and creative problem solving. The game simulates electronic circuits. The editor feature allows players to build simulations.

Curriculum

Davidson & Associates, Inc. (3135 Kashiwa St., Torrance, CA 90505) markets *Math Blaster Plus*, which drills students on addition, subtraction, multiplication, division, fractions, decimals, and percentages. Another program they provide is *Algeblaster Plus*, which introduces algebraic concepts.

CALCULATORS

We strongly support the use of calculators in mathematics classrooms. The American public seems to fear that if children are allowed to use calculators in school they will not learn the necessary basic computational skills and will become dependent on calculators. Research shows that using calculators does not interfere with learning basic skills in mathematics. Calculators are readily available; our obligation is to teach children how to use them intelligently as a powerful tool in doing mathematics (Willoughby, 1990). "Calculators enable children to explore number ideas and patterns, to have valuable concept-development experiences, to focus on problem solving processes, and to investigate realistic applications. The thoughtful use of calculators can increase the quality of the curriculum as well as the quality of the children's learning" (NCTM Standards, 1989, p. 19).

MANIPULATIVES

Manipulatives are a useful way to vary instructional activities, provide problem solving experience, provide concrete representations of abstract ideas, and provide a basis for analyzing sensory data. They also promote active participation by pupils, provide for individual differences, and increase motivation (Reys, 1971). Asking questions based on the work with manipulatives and allowing students to make mistakes also enhances their learning.

Reys (1971) suggests six criteria for selecting manipulatives. The materials should:

(1) provide concrete representation of mathematical principles and be mathematically appropriate;
(2) clearly represent the mathematical concept;
(3) be motivating (e.g., attractive and simple);
(4) be useful for developing more than one concept;
(5) provide a base for abstraction; and
(6) provide for individual manipulation and explore as many senses as possible.

Curriculum

The importance of studying mathematics visually and with manipulatives cannot be overemphasized. For example, it is easier to understand (and recreate) the Pythagorean Theorem visually than solely in an abstract fashion.

The manipulatives listed below are available from commercial publishers. Many of them are accompanied by books that explain how they can be used to teach mathematically sophisticated ideas.

Attribute blocks are a powerful manipulative giving young children practice with classification and much more. For example, they can be used as Venn diagrams in junior high school and later in more advanced high school mathematics such as discrete mathematics. Teachers have found that children use higher level thinking skills with the manipulatives compared to without them.

Clear Plastic Solids (from Cuisinaire) can be used to teach geometrical concepts.

Counters are small figures, about one inch tall that come in different colors. These can be used for counting, sorting, classifying, and grouping. At home, children can start a button or shell collection. Counters can be purchased from commercial publishers, but they can also be found in toy stores (e.g., parents or teachers might buy a bag of plastic animals for sorting). They can also be used for studying sets, bases, and statistics.

Cuisinaire Rods are useful not only for counting, but also for studying different bases. These are a "classic" manipulative. A key to their success is the use of color and length as their only distinguishable characteristics. Children can physically represent number relationships by assigning any number name to one of the rods. Once they do that, they can find names for the remaining rods relating to it.

Multilinks cubes can be used for classification and place value, studying geometric concepts such as area and volume, and for exploring operations on whole numbers.

Pattern Blocks are often used at the primary level for patterns, but can be used for older children who are learning about fractions. The book, *Fractions with Pattern Blocks* by Matthew E. Zullie (Palo Alto, CA: Creative Publications) provides suggestions for using them in the classroom.

Pebble Math are games and puzzles. They teach arithmetic facts, problem solving, spatial and visual perception, sets, logic, and strategy.

Curriculum

Tangrams are geometric shapes that children can put together to make designs. They are useful for copying, matching, problem solving, and logical thinking skills.

Other manipulatives include geoboards, base ten blocks, colored squares, and fraction bars.

A popular program that makes good use of manipulatives is *Math Their Way*. This comes highly recommended by preschool, primary, and upper elementary teachers. This program is probably not suitable for use at home because it is rather expensive. Other sources of manipulatives include publishing companies such as Creative Publications, Cuisinaire, IDEAL, ETA, and NASCO.

MATHEMATICS GAMES

Playing common games can also provide opportunities to learn mathematical concepts. Kamii (1985) encourages game playing and the use of dice to learn mathematical concepts. Children can develop strategies for adding and multiplying mentally. One extension of the game playing is to ask the children to write down their thought processes. Some common games that help students improve their mathematical skills are listed below.

Battleship. Two players try to sink each other's ships. The game improves memory and planning skills.

Dominoes can be used for number recognition, counting, and matching, as well as games and puzzles. They reinforce arithmetic skills and logical thinking.

Master Mind is a logic game for two people. One person sets up a code with colored pegs while the other tries to discover the code by making guesses.

Othello. Elementary children can easily learn how to play it, but it's absorbing even for adults. This is a game of strategy for two players that improves spatial skills.

Pente is a logic game for two people. The rules are simple, but the game can involve complex thinking processes.

PERIODICALS FOR STUDENTS AND TEACHERS

Magazines such as *Discover, Games, Gifted Child Today, Mathematics Teacher, Arithmetic Teacher,* and *Quantum* include puzzles, games, and ideas for enriching the mathematics curriculum for talented students. More information about these resources can be found in Appendix D.

CONTESTS

Participating in contests offers students a chance to improve their skills, study a subject in more depth than they typically would, receive recognition for achievements, and travel (Dauber, 1988). Perhaps the greatest benefit of participation in contests is the opportunity to meet other students with similar abilities and interests.

Mathematically talented youths should consider every opportunity to hone their talents in competitions, from the simplest ones in grade school and MathCounts in junior high school to striving to become a member of the United States team in the annual International Mathematical Olympiad, and other major international events. Participation in these contests offers students an enriching opportunity to develop their abilities. Students can study and understand their subject at a deeper level than is typical. Mathematics contests and other contests of interest to mathematically talented youth are listed in Appendix D.

CONCLUSION

In the typical elementary classroom, mathematics instruction emphasizes basic skills, repetition, and drill. The curriculum for mathematically talented students ranges from doing nothing different to tailoring the curriculum to fit the child's needs. Depending upon the level of mathematical talent, one of four models for differentiating the mathematics curriculum can be employed. The *Enrichment Model* is ideal for above average students who are not exceptionally talented in mathematics. The *Breadth/Depth Model* permits the regular classroom teacher to differentiate for a heterogeneous group. The *Topics Model* can be used for a homogeneous group of mathematically talented students. The *Acceleration Model*, which emphasizes individualized Diagnostic Testing and Prescriptive Instruction (DT→PI), tailors the instruction to meet the needs of the extraordinarily talented student in a systematic fashion. Although the Enrichment, Breadth/Depth, and Topics Models provide stimulation for talented students, the Acceleration Model (DT→PI) is the only model that ensures a systematic progression at an appropriately challenging pace.

According to Wheatley (1988), important components of a mathematics curriculum for the gifted include:

(1) an articulated K-12 mathematics sequence,
(2) self-contained classes for the gifted at all grade levels,
(3) encouragement of methods that don't require a teacher to show a student how to discover a concept, and
(4) a formal Algebra I course no later than grade 8.

When designing a curriculum for mathematically talented youngsters, parents, teachers, and mentors should take advantage of the many fine materials that are already available, although this requires collecting from many different sources and organizing the materials into a semi-structured program with goals and objectives.

> Paradoxically, the expressed need for materials is a case of starving in the midst of plenty. There are many materials for enrichment . . . that are not used. This is because it is too time consuming for classroom teachers to sift through the many kits, volumes, and catalogs to find appropriate material for the needy student, and for the resource person to comb the Scope and Sequence to decide where the enrichment materials fit. (Sirr, 1984, p. 43)

We hope that the lists of materials provided here will help classroom teachers and mentors work more effectively with their talented students.

Curriculum

CASE STUDIES OF TALENTED YOUTHS

—BY EINSTEIN'S MUSTACHE!!! I DON'T HAVE THAT "ANTENNA FEELING" ANYMORE!

The parents of mathematically talented students with whom we work take their role as advocates for their children very seriously. In most cases, the parents are the primary advocates for these children, although sometimes a child's teacher is the first to contact us. Most of the time, the mother or father calls on us after first pursuing a variety of leads. Parents typically start by talking with the classroom teacher concerning differentiated instruction for their talented youngster. Sometimes, parents and teachers work out a satisfactory arrangement. At other times, parents are required to go through several channels and wait a number of months before their child is appropriately challenged in school. In other cases, parents are frustrated by what they see as the lack of cooperation by their school system, and they seek assistance outside of the system. Parents may find that they must repeat this process year after year, and many are frustrated because their children lose valuable instructional time while they are waiting to be identified for special programming.

The typical family that contacts us has already sought out many resources. They might speak with local experts, find enrichment materials, purchase home computers, have children participate in correspondence courses, etc., in an effort to challenge their child mathematically. We try to help parents and school personnel assist the child by working together in a cooperative mode, although, as illustrated by some of the case studies below, this does not always work out satisfactorily.

We support the approach Colangelo and Dettmann (1985) describe as cooperative; parents and school participate in a cooperative effort on behalf of the child. In this approach, school personnel and parents view each other as partners and meet to discuss common concerns on behalf of the talented child. Parents look to the school for guidance in educating their talented child yet encourage the child to question work that does not appear challenging or meaningful. School personnel attempt to meet justified requests from parents and realize that they must be flexible in meeting the needs of gifted children. Parents and schools view the child as the one who has primary responsibility for his or her own learning, and they work together to realize their goals. In the cooperation mode (Colangelo & Dettmann, 1981), parents and school personnel assume that the most effective way in which to help students develop their abilities is through special educational programming that can be provided by the school such as honors classes, grade-skipping, and ability grouping.

When parents believe that the school is not meeting the needs of their child, they often will not hesitate to inform school personnel of their dissatisfaction. They readily seek outside advice on behalf of their children and may take the child to a specialist for testing and then, based on the test results, inform the school that their child indeed has special needs. These parents also express their frustration with the school system to the child and allow him or her to view the school as the cause of problems. They are quick to look elsewhere in the community for special opportunities to challenge their youngster (Colangelo & Dettmann, 1985).

This type of interaction, conflict, is the most difficult to handle (Colangelo & Dettmann, 1981). Parents might believe that their child needs special educational programming while school personnel maintain that the school curriculum is adequate for meeting the needs of all youngsters. In this situation, school personnel often believe that parents are pushing their youngsters, and parents do not support the school and even blame the school situation for their children's feelings of boredom. The story of Eddie, described below, illustrates a family that was involved in a school/family conflict.

When approaching school officials with a request for differentiated programming for their child, we suggest to parents that they have available as much data on their child's abilities as possible. This might include standardized test scores, reports from previous teachers, and reports from professionals such as school psychologists or university personnel. We emphasize that they should approach school personnel with positive statements such as, "What can we do together to help this child?" instead of, "What are you going to do for my child?" We remind them that parents and school officials are on the same team, and that school personnel as well as parents want the children to be challenged, receive a good education, and have positive social experiences.

The observant reader will note that, in the case studies included here, only one girl is described. We have consistently found that fewer mathematically talented girls than boys are nominated for the testing and special programs that we have provided. Because so few girls are even nominated, the ratio of boys to girls in these programs is quite high. This is an area that certainly deserves further investigation. It may be especially important to identify mathematically talented girls in elementary school, before the distractions of adolescence and the peer pressure not to be superior at math begin in earnest.

JOSH: IMPLEMENTATION OF THE DT→PI MODEL

Josh, the eldest child of three, was born in November, 1979. Though the first signs of his precocity were demonstrated in infancy--for example, he sat independently earlier than most babies--the most remarkable examples of his extraordinary talent came during the toddler months. At eighteen months he recognized letters and numbers and became enamored with them. Wherever he went, he searched for letters and numbers. By the time he was three years old he was reading first-grade-level books.

In addition to an "obsession" with letters and numbers, Josh developed an interest in time when he was two years old and could tell time with a standard clock by age 2½ years. Not only is the fact that he could tell time by 30 months of age important, but the manner in which he taught himself to tell time is equally revealing of his extreme talent. Josh taught himself to tell time on a standard clock by running from the room with the standard clock to a room with a digital clock and using the digital clock to verify his "guesses." Of course, since he could tell time with a digital clock, he could count and recognize numbers to 60, which was a simple step away from counting to 100. By three years of age he had caught on to the concepts of addition and subtraction. Josh's tenacity and persistence in learning to tell time and read at such an early age seem also to characterize his approach to dealing with challenging experiences.

Josh's mother realized that her son's abilities needed nurturing and therefore spent time looking at the curriculum and programs offered in the near-by public school. Mrs. J. was told very directly by the principal of the public school Josh would have attended for kindergarten that all children, no matter what level of achievement, would be expected to do the same work. Also, having a late fall birthday, Josh would not have been able to enter kindergarten until he was almost six years old. Consequently, Mrs. J. looked for a setting that would maximize her son's talents and found a private school that placed him in the joint kindergarten/first grade class when he was 4 years, 10 months of age. Josh thrived in this learning setting, where he was allowed to progress at his own rate. He loved his learning experiences so much that he would go to school even on days when he wasn't feeling well.

By the first semester of his second grade year he was working on mathematics geared for fourth and fifth graders. During this first semester of second grade, Josh's family moved to a different region of the country, where they presently live. This section of the country had a different age requirement for entry to kindergarten. Thus, Josh was almost a year younger than his grade-mates. After being told by the teacher and guidance counselor of the new school that working above grade level would result in a burned-out child, Mrs. J. contacted us and asked for our advice concerning whole-grade and subject-matter acceleration.

Josh was fortunate enough to attend an elementary school that allowed him to participate in the Mathematical Olympiads for Elementary Schools (MOES) when he was in fourth grade. Not only did he do very well in these competitions, but he also greatly enjoyed the challenging problems. Josh particularly enjoyed opportunities to work on stimulating computer programs with his classmates. He was also quite competent at programming with BASIC, and did many of his written assignments with a word processing program.

The students in his school were homogeneously placed in math according to their ability and Josh was in the highest group. However, Josh's fourth grade teacher used a fourth grade text, which was mostly review for Josh. Two grades earlier Josh understood how to add and subtract fractions with unlike denominators, thus a course in which the primary objective is learning how to rename fractions so that one can add and subtract was not only repetitive, but also unchallenging and unstimulating. Continued exposure to unchallenging material had resulted in Josh forming habits of overly fast, superficial responding without adequate thought and reluctance to "show his work." We have found that this is common among fast thinkers who thereby become "turned off" to the conventional classroom routine. These bad habits hinder the student's progress when they finally get more difficult but appropriately challenging material.

Despite the challenges presented by participation in the MOES and working with the computer, Josh's parents realized that their son was not being presented with a mathematics curriculum that was geared to his level. During the spring semester of Josh's fourth grade year, Josh and his family traveled to Johns Hopkins University for an educational assessment. Mr. and Mrs. J. wanted specifically to measure Josh's aptitude and achievement in math and to determine if he was ready to study algebra. Because Josh's parents were interested in a mentoring experience for Josh, the focus of the assessment was the development of recommendations that would be most useful for the mentor.

Throughout the assessment, there were several indications of Josh's sophistication in reasoning. He gave very detailed, accurate responses and demonstrated an excellent range of knowledge and in-depth perception of verbal concepts. A good example of his sophistication in verbal reasoning came when, after answering a question assessing his social awareness, he discussed the potential for crisis after the Chilean fruit scare.

Josh demonstrated superior skills, excellent concentration, little impulsivity, and a desire to do well on the Verbal subtests of the WISC-R. His scores on the Performance subtests, however, were *relatively* weaker and reflected some impulsivity, and a reluctance to check over his final product; such behaviors are often demonstrated by children who are not sufficiently challenged by the material.

Cases

As discussed in Chapter 4, the first step in the Diagnostic Testing → Prescriptive Instruction process requires that an aptitude test be administered. Josh was given the *School and College Ability Test* (SCAT), a test of basic verbal and quantitative abilities. The Intermediate Level of the SCAT, which was developed for Grade 6 Spring through Grade 9 Fall, was administered. Josh's Quantitative Reasoning Score was at the 99th percentile.

At this point we had much evidence that Josh's ability to succeed in an academic setting was clearly very superior when compared to other students his age. We needed, however, a more specific indication of his *achievement* in mathematics. The next step in the DT→PI model was the administration of out-of-level achievement tests, the *Basic Concepts* and *Computation* subtests of the *Sequential Tests of Educational Progress* (STEP).

In the DT→PI, Level 4 (normed for 3rd, 4th, 5th, and 6th graders) of the STEP is the first level administered. Josh completed both the *Computation* and *Basic Concepts* tests of Level 4 in half the allowed time. He answered all of the *Basic Concepts* questions correctly, and missed two on the *Computation* section. Josh's performance on Level 4 still did not provide the information concerning what topics Josh needed to study, since he had mastered those, but we were able to gain information from the way Josh took the tests and the fact that he made two errors on the *Computation* test. The pattern of making more errors on the *Computation* test as compared to the *Basic Concepts* test is not unusual; like many talented children Josh had become bored by computation tasks that he had mastered, and preferred to work problems in his head or to guess at an answer instead of working them out on paper and checking his results.

The next step in applying the DT→PI process to Josh's case was the administration of the STEP *Basic Concepts* and *Computation* Level 3. He missed items in the areas of probability, algebraic relations, and geometry/logic; thus, we had a point at which to begin his lessons with his mentor.

Finding an appropriate mentor, and getting the school to cooperate were the next challenges faced by Josh's parents. For this case the ideal seemed to be a very talented high school teacher who had an excellent background in mathematics. Josh's parents and school principal met at the end of the fourth grade year and decided that in lieu of participating in the regular math class, Josh would be able to work on mentor-assigned homework. In addition to the recommendation that Josh work with a mentor, we strongly recommended that Josh continue to participate in contests such as the Mathematical Olympiads for Elementary Schools as this type of setting was ideal for interacting mathematically and socially with other talented students.

Josh spent his fifth grade year thoroughly mastering the pre-algebra curriculum. He was ready for algebra when he entered sixth grade, which he successfully completed earning A's and "high honors." His school's math team was selected for MathCounts. His team came in ninth in the state competition and Josh finished in tenth place individually. Josh was ready to participate in one of the regional Talent Searches by taking the College Board Scholastic Aptitude Test (SAT). At the age of 12, Josh earned a 700 on the SAT-Mathematics and a 550 on the SAT-Verbal. These superb scores are further validation of the need to continuously challenge him in mathematics.

MATTHEW: LONG TERM APPLICATION OF THE DT→PI MODEL

Matthew entered a mathematics mentor program at the age of 10, in the early spring of his fifth grade year. His parents found that it had become more and more difficult with each passing year to work out accommodations with schools concerning his math education. In first through third grade, he was enrolled in a school with supportive personnel who voluntarily provided accelerated math instruction for Matthew and many other mathematically talented students. In fourth grade Matthew and his family moved to a new state and had to fight bitterly in order to reinstate accelerated instruction.

In the middle of Matthew's fifth grade year the family moved again to another state. In his previous school, Matthew had been studying sixth grade mathematics. The principal at the new elementary school looked at Matthew's records and his parents' request for continued math instruction at the current level of acceleration (one grade level), and flipped the file shut. "Well," he said, "Looks like Matt's going to be ahead of his classmates. We just don't do that sort of thing."

On her own initiative, Matthew's classroom teacher obtained materials and continued Matthew's accelerated math so that he would be challenged in her class. Matthew's parents note that they have found that many times an individual will be accommodating, even when the institution cannot or will not.

Shortly thereafter, the gifted education teacher in Matthew's school reported that she had learned of a new program at a local university for upper elementary students gifted in math. The parents contacted the program administrator and scheduled an appointment for testing, which took place on the college campus. Prior to the test administration, Matthew and the examiner talked about the testing procedures and the mentor program itself.

It was explained that the decision to enter the mentor program would be primarily Matthew's. Since participation in the program would require time outside of his regular school day and homework in addition to his regular schoolwork, it would require his interest and commitment. The program's goal is to reach those students who are internally driven by a need to learn mathematics. Care is taken when testing and speaking to families to discern whether the interest in math and desire to learn more comes from the student or from a parent who feels that his or her child is not being 'challenged' enough in school.

Matthew took a series of aptitude and achievement tests and subsequently qualified for participation in the mentor program. This information indicated that he had mastered much of what 8th graders study in mathematics. Clearly, Matthew is a student who is *exceptionally* talented in mathematics.

Based on the results of this diagnostic testing, Matthew was matched with a mentor. The mentor assigned by the program was "Mr. B.," a doctoral student in mathematics about to embark on a career as a mathematics professor. Mr. B. and Matthew both hit it off right away. Matthew met with him for two hours a week, for which his parents paid an hourly tutoring fee. During that time, Matthew and Mr. B. covered the material to be taught, outlined assignments, and worked out any problems Matthew had in understanding previous materials.

Later, Matthew admitted that he had been afraid that he would be asked to do things that he didn't know how to, and be embarrassed if he couldn't come up with the answers. Mr. B. also noted that he had some concerns about the mentoring sessions. Matthew soon learned that he would indeed be asked to work difficult problems, but that Mr. B. would be there to instruct and encourage him. Matthew was given Mr. B.'s home telephone number and told to use it if he ran into problems. The first telephone call was agonizing for Matthew. He was literally unable to dial, because he just couldn't believe that a teacher actually meant it when he said he could "call for help." He learned that Mr. B. really did want Matthew to call. With a week between sessions, it's important for the student to know that he can reach the mentor with questions, and not remain stuck in one spot.

Matthew's enthusiasm for math, always high, was now totally set loose. He was being asked to work harder and longer on the subject than he ever had before. On the average, he spent an hour a day on math homework during the week, plus the two hour mentoring session. At first it took some adjustment on Matthew's part to learn how to plan his time efficiently. He was also dealing with the demands of coping with a new school, and had just joined a local community orchestra. Fitting new friends and Nintendo in on top of all that got a bit tricky at times, but after about three weeks he fell into a workable schedule.

After they gathered up speed, Mr. B. and Matthew covered nearly a chapter a week in the textbook. Although Matthew's time investment into mathematics had increased over that spent even in regular accelerated classroom work, he was not resentful of that fact. His efforts were no longer being spent in endless repetition of problems that he already knew how to do. For example, Mr. B. would often assign every third or every fifth problem in a chapter. Once when his mother asked how many problems he had to do in a certain section, Matthew's answer was "Mr. B. told me to do every third one in this section until I understand it. Then I can go on."

Mastery of the subject had become the main objective, and Matthew had to be able to demonstrate that mastery. He was expected to always have his assignments done and also took periodic assessment tests. Sometimes a test would show that Matthew needed more work to fully understand a certain concept, and they would go over it again, probably from a different direction. The tests and assessments were used to determine the appropriate time to move on to new materials--not for grades. The prize was not a grade, but the opportunity to move on to the next step. For students like Matthew, who are driven from within to know, knowledge truly is its own reward.

WHAT HAPPENS REGARDING MATH IN THE SCHOOL WHEN A CHILD ENTERS A MENTOR-PACED PROGRAM?

There is a lot of variation on this point, and it is best negotiated with each individual school. Matthew's elementary school principal thought it logical that, since math was so easy for Matthew, he shouldn't mind repeating the fifth grade curriculum while also doing the mentor-assigned work. His explanation was that since Matthew was currently in the fifth grade, the state needed a grade for fifth grade

Cases

math filled in on Matthew's report card. Mrs. M.'s suggestion that they use the grade from the previous year, when Matthew had already completed a fifth grade math curriculum, was not well received. The school personnel and Matthew's family eventually came to an acceptable compromise, whereby Matthew had four days a week to do his daily mentor assignments, and the fifth day belonged to the classroom teacher to give weekly tests for a school math grade. Matthew was well satisfied with this arrangement, and grateful for the time in class.

In the months they spent together, Mr. B. and Matthew covered the transitional math textbook used in the middle school. They worked on a less regular basis throughout the summer. Late in the summer, Matthew took additional mathematics achievement tests, and once again, he demonstrated mastery of the material he had been studying.

After completing the pre-algebra curriculum with his mentor, the plan was for Matthew to re-enter the school mathematics curriculum. He had completed fifth grade and was ready to move from the elementary school building to the middle school.

Matthew was ready to begin Algebra, which in his school system was not regularly taught in middle school but was offered as an eighth grade Honors class. His parents were, in effect, asking for the school to let Matthew bypass three years of their mathematics instruction (grades 6, 7, and 8). Mr. B. and Mrs. M. met with the middle school administrators. They took full records of Matthew's acceleration in math up to that point and documentation of Matthew's abilities, as charted by the standardized testing conducted by university personnel. Mr. B. spoke eloquently as an advocate for Matthew.

The middle school personnel were totally supportive. Due to scheduling conflicts they could not put Matthew in a class with eighth graders, so they proposed that Matthew meet daily with the Algebra I teacher in the resource room. That proposal worked out very well. Matthew did the same assignments and took the same tests as students in the regular classes taught by the Algebra I teacher. He enjoyed the one-on-one relationship with his teacher, who was excited and supportive regarding this new arrangement. They became good friends, too. There was no ambiguity about grades because he was taught standard materials by a regular classroom teacher. The school gave him whatever grade he earned. Matthew routinely scored higher than any student in the regular math class on assignments and exams. The middle school administrators took the initiative to see that Matthew would be provided with high school credit for the work he completed while in the middle school.

Midway through the school year, about a week before his 11th birthday, Matthew took the Scholastic Aptitude Test. He scored 690 (91st percentile when compared to college-bound high school seniors) on the mathematics section and 460 (61st percentile) on the verbal section. Clearly, this young man is exceptionally talented in mathematics and needs and deserves the special types of programs that he has been fortunate enough to have available to him. That summer, he participated in a fast-paced residential program and studied Geometry. When he returned to school in the fall, he studied Algebra II one-on-one with the same mathematics teacher he had the year before.

Matthew's parents note that they have learned to take his education one year at a time, although they are planning carefully for the future. His teacher has discussed with them the course of study she is planning for Matthew's next two years in middle school. When he reaches high school, he will complete the math sequence there. When he runs out of available classes, as he is apt to do about his junior year, he can continue his studies at a major college campus that is just a few blocks from home.

Matthew's mother says, "As a parent of a gifted child, I have had to continually struggle with the arbitrary and stubborn barriers presented by a standard curriculum. Our initial inquiries for change have sometimes been met with resistance and hostility. I have been told "He's NOT the highest IQ we've ever had at this school," as though the school is supposed to pick one bright student who is entitled to an education. I have been told "We're not equipped for that sort of thing," as though they're dealing with a tornado drill. I have been asked, "What do we tell the other parents who will come in here and ask for the same treatment?" The most difficult was the district math coordinator who, as we pursued the issue of an accelerated curriculum for Matthew, painted a grim and despondent picture of all the problems--social, academic, and personal--that were guaranteed for the future. I tried to assure him that most information showed academic acceleration to be to the child's advantage and that previous generations of Matthew's family had done things out of the ordinary and managed not to self-destruct, but he brushed my reassurances away. "Why," he pressed, "would you do such a thing? Why would you want this for your child?" "Because," I answered, "it makes him happy."

Cases

Matthew's mother continued, "Mathematical thinking is central to the way he explores and thinks about the world. To place him in grade level math instruction and ask him to wait the twelve years until college for things to get interesting, would be beyond what his patience could bear.

"Our satisfaction with Matthew's participation in the mentor program lies not in any ability it may have to turn out mathematicians, but in the serendipitous effect that taking care of academic needs has on the whole child. Matthew is productive and contented, secure in knowing that he is not likely to have to repeat any academic material. This had sometimes been a major worry and frustration to him. His relationship with Mr. B. has been enjoyable and also provided an example of what professionals in the world of mathematics do on a daily basis. His mentorship was especially rewarding--Mr. B. holds a special place among all the teachers Matthew has ever had.

"What is perhaps surprising is that Matthew doesn't necessarily plan on becoming a mathematician when he grows up. His career will come as a calling to him or perhaps it will evolve over time. As parents we are grateful that when he reaches that point of decision in his life, he has a good chance of arriving at it not only with the necessary academic tools but with the one element that will give him the most enjoyment, and which [the mentor program] has done so much to preserve his joyous and unquenchable intellectual thirst."

JEFF: A BRIGHT STUDENT WHO DOESN'T NEED AN INDIVIDUALIZED PROGRAM

Jeff came to our attention when his mother read about a local search for talented elementary students. He qualified for the "search" because his scores were in the top 3 percent on the Iowa Tests of Basic Skills. He took the Talent Search out-of-level aptitude test, but *did not* score above the 50th percentile on the out-of-level aptitude test. Therefore, Jeff was not recommended for a mentoring approach to mathematics.

Jeff is like many bright fifth grade students. His high IQ (155 Full Scale on the Wechsler Intelligence Scale for Children-Revised) is reflected in his inquisitivness, and generally high scores on achievement tests such as the Iowa Tests of Basic Skills. Jeff is the "overall bright" child, i.e., he is equally strong in writing, reading, science, and social studies. He loves going to school, and is an avid participant in class discussions. However, Jeff differs from the other students in this chapter because of his *relative* lack of drive. While Jeff might be able to spend hours in front of Nintendo, and knows intricate details of every sport imaginable, his drive to do "more than the minimum" in an academic setting is absent. This is most likely due to the lack of challenge he has received throughout his years in elementary school.

Jeff's behavior is in contrast to that of students like Josh and Matthew who view mathematics as a recreational activity, and have a tremendous drive to do math. When given the option to do math or play basketball, Jeff would not hesitate to play basketball. Josh and Matthew, on the other hand, would choose math.

Even though Jeff and Josh earned IQ and abstract reasoning scores that were identical, these scores are not enough to indicate individual learning needs. Jeff differs from Josh and Matthew in motivation and persistence, however, he is similar to them in needing more challenge than regular, grade-level math. Jeff needs *enrichment*, but not the typical worksheet type of enrichment usually found in math classes. Jeff enjoys the challenge of working with computers and of problem solving and needs to have the stimulation of participating in competitions such as MOES and MathCounts.

It's easy for students like Jeff to slip through the cracks and remain unchallenged and unstimulated. These students often develop poor study habits and poor skills for problem solving. Inconsistency in school performance can result in serious underachievement in high school and college. We strongly encouraged Jeff's parents to advocate for him by seeking challenging experiences and helping him find ways to stretch his mind.

EDDIE: IN THE MIDST OF A PARENT/SCHOOL CONFLICT

During his fourth grade year, Eddie expressed boredom in school so his parents asked school officials to have him tested to determine what his academic needs were and to place him in the proper program. He took two preliminary tests: the Slosson Intelligence Test, on which he earned an IQ of 137, and the Key Math Test, on which he earned a score placing him one year and five months above his age peers. According to school personnel, the results of the testing indicated that he was eligible to be tested for admittance to the

Cases

Academically Talented (AT) program. Unfortunately, an entire year passed before he was tested for that program. The AT program at Eddie's school is an enrichment program, and requires that all of the classroom work be completed, in addition to the enrichment work. Mrs. E. was concerned that, "Even if he is admitted, this does not mean he will receive advanced training in mathematics or science."

When Eddie was in fifth grade, his parents reported that he seemed bored and didn't perform as well as he could. He was doing the same work as the other 5th graders: three digit addition and subtraction, two digit multiplication, and long division. He did well on tests and on work on a one-to-one basis, but he was careless with routine tasks such as those found in typical homework assignments. Eddie's teachers commented that he was not attentive to detail and needed work in improving his attention to tasks at hand. He earned the following grades during the first marking period in 5th grade: B in reading, B in math, B in English, C in Spelling, and A in Social Studies.

Although Eddie was unchallenged in school, he was not idle at home. He and his parents found a number of interesting things for him to do that challenged him mathematically. For example, Eddie enjoyed working on the computer at home. He used *The Science Toolkit* computer program published by Broderbund and worked on other science related projects. He also used the *Algeblaster* software, which reinforces concepts for pre-algebra and algebra courses. Eddie quickly caught on to the ideas presented in this software with little help from his parents and no supporting text. Eddie also tried Borenson's Hands-On Equations Learning System, and again quickly mastered the concepts without much parental guidance. These activities were completed during the summer before his fifth grade year. When his parents contacted us in the fall of his fifth grade year, Eddie was taking an Algebra I correspondence course in a university program using Harper & Row's *Algebra I* text. He did not have difficulty with the correspondence course, but the work went slowly because he was doing it in addition to his regular school work.

Our first step in developing a mathematics program appropriate for Eddie was to document his progress and achievement in mathematics with standardized mathematics achievement tests. At the age of 10, he was at the 98th percentile when compared to other fifth graders.

After Eddie completed those tests, his parents wrote, "We are open to any suggestions. We have tried to work with the school's administration without any success. We asked to have Eddie taught math at the intermediate school but were turned down because 'it was a scheduling problem' for them. Eddie is, needless to say, still quite bored with most of his school work. He enjoys challenges, and is looking forward to your next test. . . . We will continue to fight to get him a challenging education to keep his desire for learning strong. We feel that your evaluation will carry much weight with the school administration since our district participates in the SMPY program at higher grade levels."

Next, Eddie completed a more difficult level of the mathematics achievement test and earned a score at 87th percentile when compared to first semester 7th graders. After the standardized testing was completed, Eddie's parents met with elementary school officials. The school decided to let Eddie take a test that covered the rest of the 5th grade math, stating that, if he did well, they would make accommodations for him. His parents attributed this change in policy to "the independent testing and the fact that there was also recently a court case in [our state] in which a school was forced to offer a child an appropriate math education."

Although all of this sounded promising, Eddie was soon in the midst of a school/home conflict. School officials apparently were opposed to accelerating students and also were hesitant to make special accommodations for Eddie because of his poor work habits. His parents were upset because they had been asking school personnel for help for a long period of time and weren't making much progress. According to school officials, Eddie was not demonstrating the same level of achievement in mathematics at school as he was at home.

To determine Eddie's level of mathematics achievement, teachers in Eddie's school developed a test for him that covered 5th grade math. Eddie had a difficult experience while taking that test; it was administered over a five-day period during a time when he was especially affected by his hay-fever allergies. After Eddie took this test, Eddie's mother reported that the "school is hostile." She was extremely frustrated because of the roadblocks that she perceived school officials had put in the way of Eddie's progress.

Nevertheless, Eddie did perform well enough on the 5th grade mathematics achievement test developed by his teachers that his school hired a teacher to work one-on-one with him beginning in February of his fifth grade year. Unfortunately, Eddie was not completely happy. His mother reported that he felt pressured to hurry through the material; they were half-way through the 7th grade

Cases

textbook after 10 meetings. There was no communication between the previous math teacher and the new one. School officials also had not decided whether or not he would be given credit for the work.

Soon after Eddie's mother talked to us, the gifted/talented coordinator at his school called. The coordinator reported that Eddie had poor work habits: his work was messy, was turned in late, etc. She was frustrated because Eddie didn't pay attention to the teacher and frequently got up out of his chair and walked around the classroom. He also had gaps in his background; for example, he didn't know his multiplication tables, couldn't multiply fractions, and didn't know decimals. Another concern she had was that he worked quickly but with many errors.

During the past few years, Eddie had been absent from school 20 percent of the time because of his allergies. The coordinator commented that sometimes he used these allergies as a crutch or to manipulate adults. For example, he might say, "I can't do this--I feel an asthma attack coming on." The math tutor at school commented that when he encountered difficulties solving math problems, he would find an excuse not to do them. Eddie had learned to manipulate adults by using his allergies or other health problems as an excuse. Based on their observations, school personnel believed that his mother had an inaccurate perception of Eddie's abilities. His mother countered with the fact that his condition was medically documented by an allergist and a pediatrician.

While all of this was happening at school, Eddie's family had contacted the mathematics department of a nearby university and found a graduate student who was willing to serve as Eddie's mathematics mentor outside of school. The mentor, "Charles," also tutored other students, although Eddie was the youngest student with whom he had worked. Eddie and Charles met outside of school time, and Eddie's parents paid Charles a tutoring fee.

Eddie and Charles soon developed a good rapport. Eddie's mother wrote, "Up until this time, we never realized how long something could hold Eddie's attention. He's always had a short attention span at school and this has driven some of his teachers up the wall from time to time. Now, however, he spends 1½ to 2 hours twice a week with Charles and comes out of the sessions very excited and happy. . . . Charles has never complained of an attention problem."

The standardized mathematics tests described above had been administered by Eddie's parents at home. School officials were concerned because parents supervised the testing and requested that one of the teachers or the principal administer tests in the future. Eddie's mother responded that, since the school was not paying for the tutoring Charles provided and since they were not giving Eddie credit for his work, she didn't want the school principal to administer the tests. "We view this [work with the mentor] as enrichment and Eddie knows that if he doesn't want to go he doesn't have to. The people at school have in the past put a lot of pressure on Eddie in test taking situations and we refuse to let them do it again. We did agree to have some more testing done but we refused to let the school personnel perform it."

Eddie's parents were militantly fighting for what they perceived he needed for his education and resisted many suggestions the school made. We reminded them that helping Eddie to get the best, most appropriate education was their goal; it was the school's goal and our goal as well. We encouraged them to recognize Eddie's difficulties with some basics of mathematics, such as multiplication facts. Eddie needed to move ahead at an appropriate pace, but also he needed to learn basic facts. We reminded Eddie and his parents that no matter how boring it might be to study multiplication tables, it is necessary to know them.

Eddie's poor work habits may have resulted from lack of challenge; however, we encouraged Eddie to try harder to be attentive and to respect the classroom rules. Eddie's mentor had noted that he was attentive during mentoring sessions, which indicated that Eddie could control his behavior.

Eddie's parents and school were in conflict, rather than working cooperatively on Eddie's behalf. Unfortunately, Eddie got caught in the middle. Our goal was to work cooperatively with the school and parents in order to help Eddie in appropriate ways.

Eddie's school tutor began setting weekly goals for him. For example, one week his goals were to turn in all homework, get a grade of 90% or better on all quizzes on multiplication tables, and to pay attention in class. Eddie's mother noted that his health had improved; he had missed only 3 days of school between January and mid-April. He continued to work with his math mentor outside of school, too.

In one letter, she wrote, "The DT→PI method not only makes sense but has really made a difference for Eddie. Before, he would skip easy sections of tests and only do the harder problems. Needless to say, his grades in class would be very erratic. This really hurt his self-image because he would later feel "stupid" at having missed the easy

Cases

ones despite the fact he could do the harder problems. He's not perfect but it has helped him to concentrate more because the work is generally more interesting. He knows that the math that he does with Charles is 'hard' and this has really helped him to feel good about his abilities. He was really proud of himself for having passed the school's 5th grade math test."

"Even though at this point we have no idea what will happen [next] September, we would encourage other people to strive to change the system. Eddie's teacher now has other gifted students that she is working with. So, things should improve for those behind Eddie. . . . Ultimately, parents must be responsible for their children's education and must be willing to work to overcome the status quo." "[The school] would have been content to let Eddie cruise because his grades were good despite the fact that he was bored and losing the desire to learn. Parents must be persistent and not give up as obstacles are thrown their way. Every time we see Eddie come out of a math session with Charles on a 'real high,' we know the effort was worth it."

Eddie's parents contacted us again about 18 months later. They reported that, as a fifth grader, he had completed the 7th grade math course. Sixth grade required a move to a new building for middle school. The middle school resource teacher expressed an interest in providing individualized mathematics instruction for Eddie and contacted his parents for information about his abilities and achievements. Eddie's parents expressed frustration because the elementary school personnel "had not bothered to share any of the test results or [the literature we had collected on mathematically talented youth] with the middle school administration. Again, we supplied all the materials and waited."

While they waited, Eddie discontinued his meetings with his mentor and attended a mathematics enrichment program offered by a local university. Soon after Thanksgiving vacation of Eddie's sixth grade year, the resource teacher informed Eddie's family that she would not be working with him. Instead, he began meeting one day a week with the seventh grade math teacher to work on mathematics enrichment. The rest of the week he attended the regular sixth grade math class. Eddie's parents accepted this approach even though they believed that it was not providing an appropriate math education for Eddie; they still wanted to try to work within the public school system.

Eddie's parents reached the last straw when it came time to prepare Eddie's Individualized Educational Plan (I.E.P.) for seventh grade. The principal of the middle school wanted him to repeat seventh grade math because one of the middle school teachers had not instructed Eddie in the seventh grade material. Eddie's parents were concerned that this would have a negative effect on Eddie's self-esteem. The principal did not believe Eddie had any exceptional abilities in mathematics despite the fact that he had earned the highest score in his school on a recent state math competition. Mrs. E. says, "Needless to say, after three years I gave up and started the search for a private school in our area. We left on good terms from the middle school. . . . It was a very different situation than elementary school but the same problem--Eddie did not fit the 'school model' for a talented child."

Eddie's parents removed him from the public school and enrolled him in a private school. At the new school, Eddie took some classes with age peers and but studied mathematics and German with older students. The mathematics course he took that year was "Enriched Pre-Algebra," an eighth grade honors math course. His parents reported that he had a "wonderful" seventh grade year both academically and personally. "The obvious drawback to this plan of action is, of course, the cost. We are spending his college money now, but if he loses interest in school altogether, what good would it be anyhow?" That year, he took the Scholastic Aptitude Test and scored 540 on the mathematics section and 390 on the verbal at the age of 12 years 6 months, which was high enough for him to be invited to participate in a residential summer program offered by one of the regional talent searches. He attended that summer program and completed Algebra I during the three weeks. At the end of the program, he earned a score at the 99.4th percentile for Algebra I, and it was recommended that he receive full credit for a year of Algebra I. Eddie reported that he greatly enjoyed the intellectual freedom he experienced in this program. He said, "You were told where to start and then let go at your own pace, with frequent tests to track your progress. If you needed help you could ask the teachers." Eddie also reported that he enjoyed the recreational aspects of the program. "The only thing I didn't like was the food at the cafeteria." Eddie expected to be placed in the Geometry or Enriched Geometry class upon returning to his school in the fall.

After reading a draft of this case study, Eddie's parents emphasized one thing: "If a child does not fit the school administration's criteria for 'gifted,' the child is at risk of being lost forever in the system. Ultimately, parents need to recognize their child's potential, to persistently pursue creative avenues to enhance their child's education, and to be willing to stand up for their child in public. These are not easy or popular things to do. However, if we had not taken this course, I think Eddie would never have been able to take a chance and reach for his potential. In that atmosphere, he would never have developed the self-confidence and self-esteem needed to try a great challenge at which he might ultimately fail."

In looking back on Eddie's story we noted that whatever could go wrong between Eddie's parents and the school personnel did go wrong, and Eddie was caught in the middle of the conflict between his parents and his teachers. The antagonistic relationship between Eddie's parents and the school became so intense that his parents believed that the only real alternative was to go the private school route.

The relationship between parents and school had degenerated to the point that even providing objective evidence (i.e., standardized test scores that indicate what has been learned, and where to begin further instruction) and trying to make programming decisions based upon that evidence was not possible. Eddie's parents were suspicious of the school personnel, and school personnel were suspicious of the parents motives. Unfortunately, this situation is not uncommon.

When we try to help parents in this type of situation, we attempt to do three things:

(1) help them sort out their concerns for their child's educational programming from their overall concerns about education,
(2) suggest ways of *objectively measuring* their child's aptitude and progress and provide guidance for obtaining that assessment, and
(3) give specific guidance regarding their approach when dealing with school personnel. Teachers and administrators are extraordinarily dedicated individuals. They are also very sensitive, and have a great deal of ownership for their programs. However, many parents do not realize how this dedication can lead to a sense of feeling threatened when it is suggested that a student's needs aren't being met.

ELIZABETH: A MATHEMATICALLY TALENTED GIRL WHOSE PARENTS DON'T WANT HER TO BE DIFFERENT

Elizabeth's mother initially contacted us with vague concerns about the lack of challenge her daughter was receiving in school. She was also concerned about her daughter's self-esteem. Mrs. B. was worried that Elizabeth was bored and might become a behavior problem; she was also quite concerned about placing Elizabeth in any type of program that would make her dramatically different from her classmates.

Elizabeth was in fourth grade when Mrs. B. contacted us; however, Mrs. B. indicated that she had been making inquiries about her daughter's progress for several years, and had recently discussed with school personnel the pros and cons of accelerating a grade. After a lengthy phone conversation with us, Mrs. B. realized that (1) objective information was needed before educational programming decisions could be made, and (2) it would be best if the school would monitor an educational assessment. It was suggested to Mrs. B. that she could let the school officials know that we were willing to work cooperatively with school and parents to secure challenging work for Elizabeth.

Fortunately, Elizabeth had a very talented gifted education teacher who fully realized that the enrichment approach of the one-hour a week pull-out program Elizabeth was in would not give Elizabeth the challenge she needed in mathematics. This teacher also was extremely secure in seeking outside advice. Thus, when Mrs. B. contacted her to say that some university people would help them identify Elizabeth's needs, the teacher called us that very day.

The first step in identifying Elizabeth's educational needs was to see how she compared to her gradremates. Mrs. B. had indicated that Elizabeth scored at the 99th percentile on all of the subtests of the Iowa Tests of Basic Skills, but she wanted to concentrate on identifying challenging mathematics curriculum. As soon as we received that information, we administered the quantitative sections from a practice *Secondary School Admission Test-Lower Level* (SSAT) as an out-of-level test. On the practice test, as a fourth grader, Elizabeth earned a score at the 79th percentile when compared to 6th graders.

Elizabeth was ready for the next level of the diagnostic assessment: evaluation of specific achievements in mathematics relative to older students. Level 3 of the STEP tests, which are normed for grades 5 through 8, was administered. Compared to Spring semester 6th graders, Elizabeth earned a score at the 87th percentile on the Basic Concepts test.

We have already mentioned that almost always, the basic computation skills of mathematically talented students lag behind their basic concepts. Elizabeth was no exception. Level 3 of the STEP Computation test was challenging for her. During the 40 minute time limit, she earned a score at the 25th percentile, compared to Spring semester 6th graders. On an untimed administration, she more than doubled the number of items she answered correctly. However, it was clear that she needed to solidify her computation skills.

It was recommended that she begin pre-algebra as soon as possible. In a meeting with the teachers, parents, and principal, individual goals for systematically working through the pre-algebra curriculum during fifth and sixth grade were identified.

Parental concerns about Elizabeth being "different" from her classmates were also addressed. The parents felt the need for further discussions regarding their daughter's giftedness and its impact on her as well as their family, and they were referred to a family counselor who works with families of gifted students. Elizabeth seemed unconcerned about being different from her classmates. She is like many of the girls we see: extremely talented but somewhat unaware of her abilities, and eager to do whatever will please her teacher. She was one of the few students in the pull-out program who had completed an independent study project. Another point that was brought out was the concern for being stressed by doing advanced work. We discussed with parents and school personnel the difference between stress and stretch.

Elizabeth is now in fifth grade, and will finish the pre-algebra curriculum this year. Because school personnel were involved in individualizing her mathematics program from the very beginning, they have readily accepted their responsibility for finding appropriately challenging mathematics curricula for her all the way through high school.

Participating in a DT→PI program has not only helped Elizabeth to become more focused in math and to get excited about math, but has helped other mathematically talented students in her district by successfully demonstrating a model for working with talented students in a specific area.

During fifth grade, as part of a search for talented elementary students, Elizabeth took the entire SSAT. On the Quantitative section of the SSAT, her score was at the 84th percentile; on the Verbal section of the SSAT, her score was at the 96th percentile; her SSAT Reading Comprehension score was at the 98th percentile, and her Total score was at the 99th percentile (all scores were compared to 7th graders). Her outstanding performance on the out-of-level aptitude test has reinforced for the school and the parents the need to continuously challenge her.

STEVE: RADICAL ACCELERATION OF A FIRST-GRADER

Steve and his two older siblings had been tutored since the age of three by their parents, and Steve had been reading since that time. By the time Steve was 6½ years old, he could perform algebra problems, type 50 words a minute, and write his own computer programs.

Steve had been enrolled in first grade for a few weeks, but was not receiving differentiated curriculum in any subject area; therefore, his parents withdrew him from school. Steve was being home-schooled by his mother, although his primary advocate was his father. Steve's father had also filed a formal complaint with the Department of Education because the school was not making accommodations for Steve. At that point, in an effort to avoid further litigation, and upon the recommendation of the school district's superintendent, Steve's parents contacted us.

A review of the records indicated that Steve had been evaluated previously at the ages of 5 years, 1 month, and 6 years, 5 months. Each evaluation included the administration of an individual intelligence test, and each evaluation resulted in confirmation of Steve's superior intellectual ability.

During these evaluations, academic achievement tests, designed to provide a general indication of a student's achievement in reading, mathematics, and spelling were also administered. On two tests designed to screen for achievement, Steve performed at the 7th grade level for reading, math, and spelling. The primary recommendation was that Steve's program of study should be academically stimulating while in a setting that provides appropriate socialization with his classmates. No action was taken by the school

Cases

towards meeting his academic needs. He was placed in first grade to provide for socialization.

Steve's parents, who were very determined that he receive appropriate educational challenges, were frustrated with the fact that they had evidence that their son needed more challenging materials (possibly at the 7th grade level), yet his teacher refused to make any adjustments in the first grade curriculum. Both Steve's parents and his teachers had legitimate reasons for their position regarding what was appropriate for him, unfortunately, they could not effectively communicate those reasons to each other. In addition, Steve's situation was somewhat complicated by the fact that his father had actively advocated for Steve's older brother and sister, and school officials regarded Steve's father as a "pushy parent."

When students who have superb ability to learn are tutored at home, it is sometimes believed that the parent's opinion is circumspect because parents have invested so much in their child's education. School officials indicated that they were suspicious of Steve's father's motives. We felt that it was important to diffuse those suspicions because we sensed that Steve's father had tapped into his son's strengths and had helped his son realize those strengths. After working with Steve, our reaction was that his behavior was similar to that of a mature, extremely intelligent eight or nine-year-old. His demeanor was like that of a well-behaved upper-elementary student.

On the one hand, Steve's father was advocating to have him placed in seventh grade based upon his experience in working with his son as well as the objective evidence from the screening instruments. However, the screening instruments that yielded those seventh grade scores were not developed to assist in designing programs. Of course, Steve's father didn't know that, and should not have been expected to know that. On the other hand, Steve's teacher had a pedagogical sense that he needed more challenging material, but certainly seventh grade material would be too difficult.

It was apparent that one of our goals had to be to determine Steve's achievement relative to his ability, and then convey that information to both the school personnel and Steve's parents so that Steve could receive a challenging education in the public schools.

What follows is a summary of the assessment and the recommendations that came from it. Most assessments begin with an IQ test, but we decided that it wasn't necessary to administer another IQ test, since he had already been given two within eighteen months. We did, however, want a sample of his abstract reasoning ability, so we asked Steve to complete the *Raven's Progressive Matrices* (RPM), an untimed nonverbal test of figural reasoning. He did so well on this test that we began to think of him as a highly gifted fourth grader in a six-year-old body.

Superior ability to process information and to attend to presented learning tasks is extremely rare and required careful tailoring of an individualized educational plan that would provide an optimal match between Steve's ability and achievement. We knew how well he could reason, but we didn't have enough information about his academic achievement to program effectively. The two previous evaluations were only a *screening* of his spelling, reading, and mathematics achievement. An assessment that was more diagnostic in nature was needed.

Three mathematics tests were administered before finding one that was appropriately difficult to determine where to begin his mathematics instruction. Steve did very well on the tests developed for 3rd-5th graders indicating that he had relatively few, if any, gaps in his mathematics knowledge base. The biggest concern was that he not rush too quickly into pre-algebra and algebra because he needed time to allow for the development of the necessary cognitive structures that would foster success in more abstract mathematics such as algebra. Unlike many extremely precocious students, Steve had not developed sloppy habits. He did not do all of his work in his head, rather he was careful to work out the problems on scratch paper. The biggest concern was that if he remained unchallenged, he would most likely develop poor work habits. Steve was at a critical point in his academic development.

In reading, by using a standardized, comprehensive test designed for students in grades 3, 4, and 5, it was determined that those skills measured by the cognitively "less-demanding" tasks of recognizing words and decoding them were more advanced than his *understanding* of common words and his general reading comprehension, especially his inferential comprehension. This explained why he did so extraordinarily well on the screening tests; screening tests are not designed to pick up the more advanced skills. This also explained why his teacher was reluctant to give him seventh grade materials; she knew that he could read the words, but she also knew that he did not have the maturity to draw conclusions or synthesize the material at the seventh grade level.

It seemed to us that providing reading materials (e.g., social studies, science, etc.) at an advanced third or fourth grade level would be instructionally appropriate. It was pointed out to the parents and the teachers that Steve's ability to decode the written word will continue to be far superior to his ability to comprehend for several more years. He needed time to allow underlying cognitive functions to develop and mature. He also needed the opportunity to interact with students who were at a similar level of comprehension. It was unlikely that these students would be found in the typical first grade classroom, and we started to discuss the possibility of accelerating Steve to a higher grade.

Since his math and reading comprehension skills seemed to be equally developed, it made sense to consider whole-grade, rather than subject-matter acceleration. For subjects such as science and social studies, Steve was ready to begin receiving instruction at a third, fourth, or even a fifth-grade level. The next job was to find the right classroom and teacher for this talented six-year-old. An understanding teacher who could adequately prepare his or her class to welcome a new student (who is younger, yet equally able), and who could communicate effectively with the parents was most important. After careful discussions with the superintendent, teachers, principals, and parents, it was mutually agreed to accelerate Steve to third grade in all subjects except math. It was determined that he was so advanced in math that he needed a mentor approach.

Steve had achieved a great deal through the home-schooling provided by his parents, but he needed and wanted the opportunity to interact with peers. He also needed exposure to extra-curricular activities and contests, such as spelling bees, the Mathematical Olympiads for Elementary Schools (MOES), and science projects that are typically assigned in the upper elementary grades.

We felt that it was important for Steve's parents, who genuinely enjoyed providing all of their children with enrichment experiences, to continue to do so and to be supported in those endeavors by the school district. However, it was suggested that these enrichment experiences should be focused on opportunities that are not traditionally offered in the regular, public school. For example, Steve could learn one or two foreign languages and take music lessons. Activities in sports and social groups such as cub scouts were also encouraged.

We discussed the need for long-range planning. For example when he reaches the age of 11 or 12 years, he might benefit from summer programs offered by the junior high Regional Talent Searches (see Chapter 8). He will soon be ready for high school level work, and will need continued accommodations for his academic program. Steve will probably be ready to enter college two or three years earlier than is typical.

PETER: AN EXTRAORDINARILY TALENTED STUDENT

Born seven weeks premature, Peter was hospitalized several times for intestinal problems, anemia, and poor liver function during his first three years of life. He showed an early interest in puzzles, shape boxes, and books about numbers. At about 20 months, he could complete a shape sorter with 18 shapes and enjoyed dividing the shapes into piles by color or general shape. One of his favorite books was "The Count's Counting Book," which he requested five or ten times a day for several months. At 21 months, he finally said his first word, "Yes," but he had nothing further to say until shortly before his second birthday. His family was watching the Kentucky Derby on television when he pointed to the screen and said clearly, "Nine!" as a horse with a large number nine walked by. The family immediately discovered he could point out any number up to 10 and could say about half of them, and also recognized most of the letters of the alphabet.

By the time Peter was 26 months old, his vocabulary had dramatically increased and he was speaking in full sentences. One evening, his mother was reading a book to him, and she was surprised when he pointed out the word "zoo" in the text. She then showed him a set of reading cards, and he learned all of those words by the age of 2 years, 5 months. His mother recalls, "This tiny, undernourished child (he was still wearing newborn gowns to sleep in) would appear at the foot of our bed at six in the morning and beg, 'Please, can I have a new word?' We continued to make word cards for him for the next two or three months, whenever he came upon a word he wanted to know, until we realized he was reading words we hadn't taught him."

By the age of three, Peter could read silently, rarely asking for help with a word, could count and read numbers past 1,000, and could accurately count more than 20 objects. He was able to calculate sums and differences of numbers less than 10 and showed signs of understanding multiplication. For example, at age 2 years, 7 months, he and his mother were making gingerbread men. Peter counted out three raisins for each. When he was told they were making four more cookies, he responded instantly, "Then we need 12 more raisins!"

Although his family showed little interest in music, Peter soon developed a passion for music. Before age 3, he could play nursery songs on his xylophone accurately and with good rhythm. He started singing the numbers on the xylophone keys for all the songs he knew ("1-1-5-5-6-6-5" etc.). He enjoyed playing the piano when he visited other people's houses. His parents enrolled him in a Suzuki violin program and rented his first violin on his third birthday. He made excellent progress. Shortly before he turned four, he demonstrated that he had learned to read music by singing and playing a piece he had never seen before. Soon after, his family learned that he had absolute pitch; he could name any note, or sing a named note accurately. He invented a game on his musical phone; someone would play two notes on the numbered phone, and he would sing the note whose number is the sum of those played. He was surprised that no one else he knew could get the right answers.

Peter's parents recognized that he was quite talented, and they were impressed by his excellent memory. He often repeated sections of books verbatim and had memorized the location of many streets in the city. At age 3½, he could give directions to the doctor's office, museums, airport, or friends' houses in terms of the streets to take and the turns to make, often suggesting alternate routes.

Before entering kindergarten at age 5, Peter spent the summer asking for increasingly difficult arithmetic problems and used much of his free time reading math books from the library. He taught himself how to calculate in binary and other bases. His interest in music continued to grow. While he attended kindergarten half days, he also took private violin lessons, violin group, piano lessons, and dance class. He usually practiced his music an hour or two a day, but sometimes it seemed as if he wanted to play all day. He began composing and writing down his own music at about this time. He also wrote neat and grammatical mystery stories in his kindergarten journal and completed with ease the fourth grade math problems his teacher gave him.

The summer after kindergarten, Peter attended summer school. He was placed in a third grade reading class to work on handwriting and to learn phonics rules in an organized fashion. For math, he was placed in a group of fifth and sixth graders working on word problems and a review of elementary math. He was an active participant in both classes and got along well socially in his reading class. The students in the math class tended to regard him as a "pet" rather than a friend. He did well in both classes.

The next challenge for the family was to find an appropriate school-year placement for this talented six-year-old. They found a

Cases

private, ungraded primary and secondary school that stressed the development of independence and self-directed learning and was quite flexible in allowing for individual differences. Peter's mother was also offered a part-time position teaching algebra and fundamentals of math at that school. Peter attended school four days a week from 9:00 to 1:40. He was fortunate to have a flexible schedule that allowed him plenty of time to continue his music lessons and practice.

As a six-year-old, Peter enjoyed programming on the computer in BASIC. In school, he worked on basic operation with whole numbers, decimals, and fractions. He was working from a pre-algebra text at the end of the school year. He did the same reading and writing assignments as his 8- and 9-year-old classmates, but he did the grammar and vocabulary work of the seventh and eighth graders. He won the school spelling bee for the Lower School (grades 3-6).

A few years later, when Peter was 9 years old, he came to our attention when his mother attended one of our workshops on mathematically talented youth. He was given out-of-level mathematics aptitude and achievement tests. His test results confirmed his extraordinary talent in mathematics. He demonstrated no significant gaps in pre-algebra mathematics, and an increasing understanding of algebraic principles, well beyond the level of Algebra I. Soon, he was participating in a mentor-paced program and studying Algebra II.

Before he was 10 years old, Peter had made extraordinary progress in mathematics, and was doing high school level math. We recommended that he take the Scholastic Aptitude Test (SAT) to document his progress relative to high school students (even though he was not yet 12 years old, the typical age for a Talent Search participant). Peter's performance on the SAT at the age of 10 years, 8 months was clearly superior; he scored 520 on the Verbal section (80th percentile when compared to college-bound high school seniors) and 780 on the Mathematics section (99th percentile when compared to college-bound seniors). Only a handful of students have ever scored above 700 on the SAT-M before age 12.

Peter's mother notes, "Peter's education thus far has been shaped by two complementary principles: His academic needs can be met in less class time than is required for the average student, and his serious involvement in music requires extra time for lessons and practice. He has made outstanding academic progress, as judged by test scores and awards, while attending school three to four hours a day in small, individualized classes. . . . He has also studied violin, piano, music theory and composition, viola, has been a member of the

Cases

string orchestra and the junior youth orchestra divisions of the local Youth Orchestra, and is a member of the professional Boys Choir. This degree of involvement with music would be impossible with full-time school attendance, unless time were allowed in his schedule for music practice or released time for music lessons."

The challenge continually presented to Peter and his family is to meet Peter's need for a flexible school program that allows him to study challenging material at a fast pace while also allowing the time for music lessons, practice, and performances. Peter's family spends a great deal of time searching for the appropriate educational placement for him for each upcoming year. They have tried several times to have him placed in a public school, but have not met with any success because of the amount of time students are required to be in class and the inability to individualize for such an exceptional student. They want him to have opportunities to interact with his agemates while also being challenged in music, mathematics, and other subjects. Peter's parents probably will need to make special arrangements for his academic education each year for the next several years.

CONCLUSION

The children and families with whom we have worked illustrate some important points.

(1) Parents have the right and the obligation to pursue those educational avenues that will result in a differentiated curriculum for their children.
(2) It is always necessary to have thorough and objective information regarding the child's progress.
(3) It is important to differentiate between issues that concern their child versus issues that concern the larger school system.

Because of the perseverance of the child's parents (most often the mother), the children described in these case studies received the challenging curriculum they needed. They all had some unnecessary delays; fortunately, children are resilient, and they are all progressing well.

Note: We thank the parents of these children and the children themselves for providing us with information for these case studies.

Assessment

7 GUIDELINES FOR A USEFUL EDUCATIONAL ASSESSMENT

—PITY THE TEST DOESN'T MEASURE ALL HER SKILLS..........

From *Of Children: An Introduction to Child Development*, 3rd edition, by Guy R. Lefrancois © 1980 by Wadsworth, Inc. Used by permission of the publisher. Tony Hall, artist.

OVERVIEW

Identification of exceptional mathematical talent usually occurs in conjunction with the identification of other abilities. Often, a student's exceptionality is first documented by a high score on an ability test that has been administered to a group of students as part of a school's annual testing program. Although group-administered ability tests are useful as initial screening instruments, their results cannot be used for specific programming, or for final identification for

111

Assessment

entrance into gifted programs.

The most reliable measure of general ability is obtained by administering an individual ability test. The central office of many sizable school systems is usually equipped to provide individual intelligence testing as well as aptitude and achievement testing, but parents may have to insist that an assessment be done. Otherwise, private, certified psychologists (who should usually have a Ph.D. degree) may be needed. An assessment by a private psychologist can be rather expensive, but may be worth the cost, especially when the psychologist is suitably trained and appropriately experienced, and therefore can help parents and educators develop an individualized educational plan.

In this chapter we discuss the benefits of an academic assessment and offer "consumer" guidelines for parents and educators who have students with exceptional talents. We briefly describe major ability, aptitude, and achievement tests, and discuss how their results can be used to develop programs that provide for students who are extraordinarily talented in mathematics.

INTELLIGENCE TESTS

Group tests of intelligence are used far more extensively than are individually administered intelligence tests. Two tests widely administered to groups of students are the *Otis-Lennon School Ability Test* (1989) and the *Cognitive Abilities Test* (1987). By their nature, group tests of intelligence require limited interaction between the tester and student and, therefore, give little information beyond a score. In general, scores obtained from group intelligence tests tend to be lower than those from individually administered intelligence tests (Sattler, 1982). Other than initially screening for general ability, group intelligence tests play a minor role in identifying exceptional abilities.

A standardized individually administered intelligence test is the best instrument for identifying gifted children on the basis of general ability. It is important for parents and teachers to understand the similarities and differences between the major individual intelligence tests, and so we will highlight four commonly used individual intelligence tests and briefly discuss the origins and uses of each.

STANFORD-BINET SCALES

The Binet scales have a long and fascinating heritage that began in the 1890's in Paris, France. In 1905, after years of testing

Assessment

school-aged children, Alfred Binet and Theodore Simon developed a 30-item test intended to measure judgment, comprehension, and reasoning. The Binet-Simon Scales came to the United States in 1908, but the first American version of the test did not appear until 1916. The scales began to acquire their present-day look with the 1937 revision conducted by Lewis Terman and Maud Merrill which was entitled the Stanford-Binet. In 1960, the Stanford-Binet was revised by selecting the best items from the two 1937 forms and combining them into one form entitled the Stanford-Binet (Form L-M). New norms for the Stanford-Binet (Form L-M) were published in 1973.

The Stanford-Binet (Form L-M) is an extremely reliable and valid instrument for use in predicting academic success. This instrument has been used with individuals as young as two years of age through adult. The major criticism leveled at the Stanford-Binet (Form L-M) is its heavy emphasis on verbal and rote tasks. Also, the Stanford-Binet (Form L-M) provides only one score, although a skilled examiner can obtain a profile of the child's performance by analyzing the items passed and failed. The Binetgram, which groups items into one of seven categories (language, memory, conceptual thinking, reasoning, numerical reasoning, visual-motor, and social intelligence), is sometimes useful in determining a profile of strengths and weaknesses based upon these seven categories. Such a profile, however, has minimal usage for educational programming.

The Stanford-Binet Intelligence Scale: Fourth Edition (Stanford-Binet IV) was published in 1986 after ten years of development and is the revision of the Stanford Binet (Form L-M). It covers approximately the same age range as the Stanford-Binet (Form L-M), and has maintained much continuity with the Stanford-Binet (Form L-M) by keeping many of the items the same. Instead of an age-scale format used by the Stanford-Binet (Form L-M), however, the Stanford-Binet IV is comprised of 15 subscales which yield an overall score measuring general cognitive functioning.

WECHSLER SCALES

David Wechsler was born in 1896--about the time that Alfred Binet and his colleagues were busy testing children with questions that eventually resulted in the 1905 Binet-Simon Scale. Wechsler became involved in intelligence testing as an Army private during World War I when the Army was conducting its large-scale testing program. This introduction to the measurement of intelligence resulted in Wechsler's eventually developing an intelligence test that would take into account

Assessment

factors contributing to a global concept of intelligence. Wechsler is the author of the original and revised editions of the Wechsler Preschool and Primary Scale of Intelligence (WPPSI), the Wechsler Intelligence Scale for Children (WISC), and the Wechsler Adult Intelligence Scale (WAIS). The WPPSI was published in 1967 and was revised in 1989 as the WPPSI-R. The WISC, first published in 1949, was revised and entitled the WISC-R in 1974, and revised again in 1991 and entitled the WISC-III. The WAIS was first published in 1955, with the revision (WAIS-R) completed and published in 1981.

The WPPSI-R was designed to measure the intelligence of children ages 3 to 7; the WISC-R and WISC-III were designed to measure the intelligence of children ages 6 to 16; and the WAIS-R measures the intelligence of individuals aged 16 to adult. Because of the nature of the Wechsler scales, very bright children often are not sufficiently challenged by the items on the test designed for their age, and their ability is only minimally measured. For example, the WISC-R does not have enough difficult questions for brilliant elementary students; therefore, they should be given the WAIS-R in order to have a more accurate estimate of their abilities. In other words, the tester needs to "shift to a test which is appropriate for an individual of the estimated *mental age*, rather than chronological age, of the person being tested" (Robinson & Janos, 1987, p. 40). Robinson and Janos suggest using a test designed for older students as an adjunct measure that enhances the information obtained from measuring the student's performance with the test designed for his or her age.

KAUFMAN ASSESSMENT BATTERY FOR CHILDREN (K-ABC)

The Kaufman Assessment Battery for Children (K-ABC) is designed to measure the intelligence and achievement of children ages 2½ to 12½. It defines intelligence as the ability to process information to solve unfamiliar problems, and focuses on two types of problem solving strategies: (1) Sequential Processing, where the child mentally manipulates stimuli one at a time in a stepwise manner; and (2) Simultaneous Processing, where the child must integrate many stimuli at once to come up with a solution. From an administration of the K-ABC the child earns four scores: Sequential Processing, Simultaneous Processing, Mental Processing Composite (combining Sequential and Simultaneous Processing Scores), and Achievement. The K-ABC is a useful instrument for quickly assessing a students's ability and achievement. Because of its limited age range, however, it may not have enough ceiling for extremely talented children.

Assessment

McCARTHY SCALES

The McCarthy Scales of Children's Abilities were developed by Dorothea McCarthy in 1972. Although the McCarthy Scales were designed to assess the abilities of children ages 2½ to 8½, for extremely able children older than 5 or 6 years of age psychologists recommend using the Binet or Wechsler Scales (Sattler, 1982).

WHAT DO INTELLIGENCE TESTS MEASURE?

Oftentimes, parents and teachers are apprehensive about having an intelligence test administered to a student. This apprehension usually stems from not being familiar with the nature of individual intelligence tests and from the potential for misuse of scores obtained from such a test. Whereas the above-mentioned tests each have unique features, they all try, in some fashion or another, to measure verbal comprehension, language development, attention and concentration, quantitative reasoning, perceptual organization, and visual motor coordination.

Intelligence tests usually have a balance of activities requiring verbal and nonverbal responses. Most tests include vocabulary as one way of measuring verbal comprehension. Short-term memory is often measured by asking students to repeat a series of numbers. The tests vary in their methods of assessing these abilities, however, and some measure other abilities in addition to those previously mentioned.

In our experience, most students enjoy taking an individual intelligence test. They like the individual attention and are usually eager to attempt the tasks. In many ways, the administration of such a test is a structured interview that lasts one to two hours, and much can be learned about the student during that time.

INTELLIGENCE TEST SCORES

IQs obtained from administering the Wechsler Scales, Binet Scales, K-ABC, and McCarthy Scales to students representative of an age group all have a mean (average) score of 100. The actual score from an assessment is most meaningful when it is converted to a percentile ranking, as this allows the student, parent, or educator to know how a child's performance compares with that of other children his or her age. If a child scores 100 on one of these intelligence tests, his or her percentile rank is 50. This means that the child scored higher than 50 percent of the children in the comparison group. As the child's score goes over 100, the percentile ranking increases beyond 50, indicating that the performance surpasses that of a larger percentage of the

Assessment

child's age-mates. For example, a child earning a score of 132 on the WISC-R would have a percentile ranking of 98, thus indicating that this child's performance surpasses that of 98 percent of children his or her age.

The percentile ranking can be misleading, however, for the upper range of scores. For example, all scores between 133 and 155 on the WISC-R are at the 99th percentile, yet the performance of the child who earned 133 differs remarkably from that of the child who earned 155. At that point, it becomes important to look at how the child performed on specific subtests and the pattern of those scores. A student's performance on aptitude tests is also helpful when comprehensively assessing academic strengths and weaknesses.

APTITUDE TESTS

Some mathematically talented boys and girls have much better verbal ability, mechanical ability, spatial relations ability, nonverbal reasoning ability, etc., than do others. These abilities are best measured by specific aptitude tests. Aptitude tests are distinguishable from ability (intelligence) tests mainly because of the more specific nature of their content. Some aptitude tests are designed to measure one specific aptitude, e.g., mechanical reasoning, whereas others measure multiple aptitudes. When measuring the aptitude of extraordinarily talented students, it is usually necessary to use a test that was designed for older, more advanced students. This notion, referred to as beyond-level testing, was discussed in Chapter 4.

SECONDARY SCHOOL ADMISSION TEST (SSAT)

The SSAT, which is prepared by test development specialists at Educational Testing Service (ETS) and administered by ETS, is a multiple-choice, secured test that is easily administered to large groups. The SSAT is given numerous times a year in many locations. Many of the over 40,000 youngsters who take this test annually are applying for admission to private elementary, junior high, or high schools.

There are two levels of the SSAT; the Lower Level was developed for grades 5 through 7, and the Upper Level for grades 8 through 11. The SSAT contains quantitative, verbal, and reading comprehension sections. Scores for the Lower Level of the SSAT are reported on a scale from 230 to 320 (SSAT Interpretive Guide, 1990). The SSAT reports aptitude scores, achievement correlations with other tests, and predictive Scholastic Aptitude Test score ranges.

We have used the SSAT as an identification instrument for mathematically talented students younger than age 12 (Assouline & Lupkowski, 1992; Lupkowski & Assouline, 1992). Based on the results of our investigations, we recommend that the Lower Level of the SSAT, which was designed for 5th through 7th graders, be administered to talented 3rd, 4th, and 5th graders (those students scoring at the 95th percentile or higher on a grade-level test such as the *Iowa Tests of Basic Skills*.) The SSAT is normed on a group of students who attend private schools, which is a selected population, so we recommend modifying Cohn's (1988) rule of thumb (discussed in Chapter 4) slightly for students taking the SSAT as an out-of-level test. Talented 3rd, 4th, and 5th graders should be compared to students in the norm group who are two years older. For example, talented 4th graders would be compared to students in the 6th grade norm group. We recommend the following: 3rd, 4th, and 5th grade students scoring at the 50th percentile or above compared to students two years older on the Lower Level of the SSAT-Quantitative should be considered for further diagnostic testing and individualized, fast-paced instruction. See Assouline and Lupkowski (1992) and Lupkowski and Assouline (1992) for more information regarding cut-offs; and "Talent Search Expands" (1991) for information about a regional talent search for fifth and sixth graders.

The SSAT was developed according to specifications similar to the College Board's Scholastic Aptitude Test (SAT), and therefore can be expected to measure mathematical *reasoning* ability in elementary school students (see Stanley and Benbow, 1986). Since it is not administered directly by schools, students and their parents have equal access to testing regardless of local program standards (e.g., Van Tassel-Baska, 1984).

SCHOOL AND COLLEGE ABILITIES TEST (SCAT)

The SCAT is a scholastic aptitude test developed by the Educational Testing Service (ETS) and marketed by CTB/McGraw-Hill. It was designed to measure the verbal and quantitative abilities of students in the third grade through the college freshman year and therefore is useful as a beyond-level test for talented elementary students. A student's performance on the SCAT is a helpful guideline for determining the need and direction for further assessment.

Assessment

DIFFERENTIAL APTITUDE TESTS (DAT)

One useful instrument for measuring multiple aptitudes is the Differential Aptitude Tests (DAT). This battery is designed for use in grades 7-12 and has two levels and two parallel forms. The DAT is appropriate for extremely talented students younger than age 12 because it has items that are sufficiently challenging. It may not be appropriate, however, for the very young child, especially children younger than nine years of age. The more brilliant the child, the younger he or she might take the DAT. Results obtained from this testing might be useful in educational programming, even though scores cannot be compared to norms for the youth's age group.

The DAT has eight subtests: Verbal Reasoning, Numerical Ability, Abstract Reasoning, Clerical Speed and Accuracy, Mechanical Reasoning, Space Relations, Spelling, and Language Usage. When using the DAT, an examiner may chose to use one, or any combination of tests to assess a student's strengths and weaknesses. Three DAT subtests that are particularly useful when assessing aptitudes of students already known to be talented mathematically are Abstract Reasoning, Mechanical Reasoning, and Space Relations. Results from these three subtests can help educators program relevantly for math as well as for science classes such as junior high physical science and high school physics.

BENNETT TEST OF MECHANICAL COMPREHENSION

Extraordinary aptitude to comprehend the mechanics of the surrounding environment often accompanies extraordinary mathematical talent. Identifying this aptitude will help student, parent, and teacher make more informed decisions regarding a student's educational program. For the young student, mechanical comprehension as an aptitude may be measured by the *Bennett Test of Mechanical Comprehension.* The Bennett is a test of aptitude rather than achievement, as the effects of training in a specific subject area such as physics have little bearing on scores. The Bennett was normed on students in grades 11 and 12 and adults in various occupations. Thus, for the highly able elementary and junior high school student it may be a useful beyond-level test. The format of the questions on the Bennett may be easier for young children than the format used for the DAT.

Assessment

RAVEN'S PROGRESSIVE MATRICES (RPM)

A useful test for measuring abstract reasoning is the *Raven's Progressive Matrices* (RPM). There are three versions of the RPM, including the *Coloured Progressive Matrices* (developed for primary children), the *Standard Progressive Matrices* (developed for children ages 6 through 13½), and the *Advanced Progressive Matrices* (developed for above-average adults). The RPM is a nonverbal test, meaning that completing the items is not dependent upon verbal responses to them. This is of importance when assessing the abilities of children in a cultural and gender fair manner. In general, we recommend that talented elementary students be given the Standard Progressive Matrices as a measure of their abstract reasoning ability, the ability to make comparisons and reason by analogy.

When taking the RPM, the student is presented with meaningless figures and asked to discern the nature of the pattern for each figure and complete the relations. As a measure of abstract reasoning, the RPM may be a good supplement to the verbal portions of individual intelligence tests. The RPM is a particularly useful instrument in determining the sophistication of the student's cognitive ability. A student who earns superior scores on an individual intelligence test and also on the RPM will most likely have the necessary cognitive structures for studying math at more advanced levels.

ACHIEVEMENT TESTS

Achievement tests differ from aptitude and intelligence tests by specifically measuring what a student has learned. They can be used to compare a student's progress with the progress of others, can serve as a baseline when measuring future progress, and can be used for placement into programs. Achievement tests play a critical role in helping teachers and parents plan for the student. Although achievement is measurable in all curricular areas, we focus on the measurement of achievement in math only.

SEQUENTIAL TESTS OF EDUCATIONAL PROGRESS (STEP)

The Sequential Tests of Educational Progress (STEP) is a battery of achievement tests assessing achievement in reading, English, mathematics, science, and social studies. The STEP tests differ from typical end-of-year comprehensive tests by encompassing broader, more general goals for each achievement area measured.

Assessment

Three advantages to the STEP tests are:

* There are several levels of difficulty for each of the mathematics tests.
* Two parallel forms are provided for each level.
* Mathematics computation and basic concepts are measured separately.

COMPREHENSIVE TESTING PROGRAM II (CTP-II)

CTP-II tests, which are marketed by the Educational Records Bureau, are available for grades 1 through 12. This battery of tests measures verbal and quantitative aptitude as well as achievement in vocabulary, reading comprehension, mechanics of writing, English expression, mathematics concepts, and mathematics computation. The usefulness of the CTP-II subtests as out-of-level aptitude and achievement tests has not yet been well documented. We expect the subtests to be quite useful for identifying talented students and determining their specific achievements because they are an updated version of the STEP tests (which have already been documented as useful out-of-level measures) and they are closely tied to the curriculum offered in most American schools. A revised version of the battery, the CTP-III, is scheduled to be made available in the fall of 1992.

COOPERATIVE MATHEMATICS TESTS

The Cooperative Mathematics Tests measure achievement in the major content areas of mathematics from pre-algebra through calculus. The series of Cooperative Mathematics Tests includes tests in Arithmetic, Structure of the Number System, Algebra I, Algebra II, Geometry, Trigonometry, Algebra III, Analytic Geometry, and Calculus. All of the Cooperative Mathematics Tests, except Geometry and Calculus, require 40 minutes of actual testing time.

Students who have progressed as far as the Cooperative Mathematics Test Structure of the Number System will want to consider taking the Mathematics and Verbal portions of the Scholastic Aptitude Test (SAT) and participating in the Talent Search Programs for junior high students. Services provided by Talent Search Programs are described in Chapter 8; addresses are listed in Appendix D.

CURRICULUM-BASED ASSESSMENT

Another way of determining a student's achievement is to

Assessment

assess what has been learned by delineating test items directly from the curriculum. This is referred to as curriculum-based assessment. Oftentimes the publisher of a textbook series will have done this for the teacher in the form of chapter and unit tests, or the teacher may choose to generate the test items based upon the specific curriculum being employed.

In contrast to assessments used for classification purposes, curriculum-based assessments provide a connection between what is tested and what is taught. There are two advantages to this method. First, it gives the instructor a baseline against which progress can be measured. Second, since the items are derived from the curriculum, instruction is directly linked to the results. Thus, the risk of omitting instruction in a certain area is avoided; the instructor and student need not worry about gaps in the student's background.

OBTAINING MORE THAN THE SCORE

A student's test score has little meaning if it is not referenced to scores obtained by other students. Scores earned by a comparative or normative group are called norms. The most useful comparative score is a percentile ranking. This number tells what proportion of the student's age-mates scored lower than the student. For example, a student earning a score at the 50th percentile performed better than 50 percent of his or her age-mates in the norm sample. The 50th percentile represents an average score.

When comparing a talented student's performance to the normative group, it is critical that the most rigorous norms be used. For example, if a seven-year-old takes the level of the STEP Mathematics Basic Concepts Test designed for 3rd through 6th graders, and correctly answers 29 of the 50 items, that score would be at the 92nd percentile when compared with 3rd graders; however, when compared to sixth graders, the same score is at the 50th percentile. In this case, the sixth grade norms should be used. The student's score is compared with norms for the higher grade level because the purpose of testing is to measure mastery of the material as well as to document the student's *advanced* achievement when compared to *advanced* students. Her score at the 50th percentile when compared to sixth graders, is an indication that material designed for sixth graders is appropriate for her.

With all tests, one can derive much useful information by closely attending to the manner in which the student approaches the test-taking task and carefully analyzing the pattern of correct and

Assessment

incorrect answers. For example, on an administration of the Bennett Mechanical Comprehension Test as a beyond-level test, one student demonstrated his anxiety in test-taking situations by rushing through the test to "get it over with" as soon as possible. When looking at this student's pattern of incorrect answers, it was noted that 57% of them occurred in the first half of the test. Most likely, he was quite anxious at the beginning, but as he began to feel more comfortable with the test he responded less impulsively and missed fewer items. Unfortunately, this student did not take advantage of the opportunity to check over his work, and needlessly missed items due to a weak strategy in which haste was the primary motivator.

The assessment should be useful immediately for planning the educational program, and should also provide information that will be helpful in long-term planning, such as when considering the possibility of whole-grade acceleration. The examiner can make important initial observations concerning the student's social and emotional development as well as the development of fine and gross motor skills.

Parents and educators might question the usefulness of a score from an intelligence test. Because of the mystique that surrounds IQs, many people do not realize that these scores simply reflect a sample of behavior that was obtained in a systematic, standardized fashion. Nevertheless, the individual intelligence test is still the most reliable and valid instrument for predicting general academic success.

THE PSYCHOEDUCATIONAL REPORT

A psychologist's report provides official documentation of the results of an assessment. In addition to reporting test scores a report should reflect the psychologist's interpretation of the child's behavior during the educational assessment in conjunction with information provided by the parent and teacher. All reports should serve as a means of informative communication. For example, a two-sentence document that states, "Johnny earned an IQ score of 145 on the WISC-R. This score indicates that Johnny is gifted," is not an informative report. At the other extreme, a ten-page document that goes into great detail with irrelevant background information and makes unfounded predictions concerning the child's future is equally unsuitable.

The initial reason for referral to the psychologist dictates the type of assessment, which, in turn, guides the information included in the report. During an assessment, the psychologist should spend an adequate amount of time getting to know the child and parents and

Assessment

thus developing an appreciation for the child's history as it pertains to the present situation. Scores from intelligence tests are often regarded as end-products, when, in fact, they are a means to the ultimate goal of developing recommendations that will guide the child's educational program.

During the assessment, the skilled psychologist will carefully observe how the child approaches different types of problems, attend to the level of language used by the child, look for indicators of impulsivity and stress, and interpret the child's general attitude in that type of setting. This information should be used anecdotally in discussing recommendations for the educational program.

The section of the report labeled "Recommendations" is actually the crux of the report. Based upon relevant background information, behavioral observations, and scores from the testing, a plan of action is developed. This plan should include a description of the educational setting and materials that will be necessary for the student's success. It should always include a method for follow-up, which does not necessarily need to be conducted by the psychologist. However, the psychologist should aid the teacher and parent in identifying the appropriate personnel for follow-up. Table 6.1 shows an example of a psychoeducational report generated after the assessment of a mathematically talented child.

Table 6.1

Psychological Interpretive Report

Student: Mark (Birthdate: 11/2/--)
Age: 9 years, 6 months
Grade: 4th

REASON FOR REFERRAL:
. . . Because Mr. and Mrs. M. are interested in a mathematics mentoring experience for Mark, the focus of the assessment was the development of recommendations that would be most useful for the mentor.

BACKGROUND INFORMATION:
Mark is the eldest child of Mr. and Mrs. M. There are two other children in the family, a daughter, age 7, and a son, age 5. Mark is presently in the fourth grade in a public elementary school in the northeastern United States. Students are homogeneously grouped for language arts and math, and Mark is in the highest ability group for both of these areas. He enrolled in this school during the fall semester of his second grade year when the family moved into the area from the southwestern U.S. Before they moved, Mark was enrolled in a private school; at this private school the curriculum was individually paced. Mrs. M. reported that in the private school Mark was working with

Assessment

Table 6.1 continued

mathematics materials geared for fourth and fifth grade students.

BEHAVIORAL OBSERVATIONS:

At first, Mark was obviously apprehensive and demonstrated nervous behavior, e.g., moving around in his chair and talking rapidly. Mark said that he enjoys writing computer programs with BASIC and making Tangrams on the computer. He was very articulate when discussing these activities. He also indicated that he is good with Microsoft Works and uses this word processing program for his language arts assignments.

As soon as we began the assessment, Mark seemed notably to relax and focus on the tasks presented. Throughout the assessment, he seemed to enjoy being challenged and demonstrated excellent concentration.

INTERPRETATION OF RESULTS:

Mark was first asked to complete the *Raven's Progressive Matrices* (RPM), an untimed nonverbal test of reasoning. For this test, the individual is presented with 60 meaningless figures and is asked to discern the nature of the pattern for each figure and complete the relations. Mark correctly completed 50 out of 60 figures and earned a score surpassing 99% of students his age in the norm sample. The raw score of 50 is equivalent to the 99th percentile for 10½-year-olds in the normed sample. Mark completed the RPM in 35 minutes and demonstrated excellent concentration.

The *Wechsler Intelligence Scale for Children-Revised* (WISC-R) was also administered. Mark earned a Full Scale score of 147, which is at the 99.91st percentile and in the Very Superior range of ability. There is a 95% chance that Mark's true score falls within the range of 141 to 153. His Verbal Score of 155 is equivalent to the 99.99th percentile. Mark earned a Performance Score of 130, which is at the 98th percentile.

The Average Standard Score for each of the WISC-R subtests is 10. For the Verbal subtests Information, Similarities, Arithmetic, and Vocabulary, Mark earned Standard Scores of 19, the highest score possible. He gave very detailed, accurate responses and demonstrated an excellent range of knowledge and in-depth perception of verbal concepts. For example, he said, "A bicycle is a two-wheeled system that when peddled moves you more efficiently than walking." On the optional subtest Digit Span, a measure of short-term memory, Mark earned a Standard Score of 12. This is a High Average score and indicates adequate concentration--it's likely that Mark's concentration might increase for more interesting tasks.

The Performance subtests of the WISC-R measure nonverbal reasoning ability. There was clearly more variability among the Performance subtests, with Standard Scores ranging from 11 to 19. The *relatively* low Standard Score of 11 on the Picture Arrangement subtest is probably a reflection of Mark's chattiness during this subtest, and consequently not receiving bonus points for completing the tasks quickly. This subtest requires individuals to sequence a series of pictures so that they tell a story that makes sense. On four of the arrangements Mark made sequencing errors; he tried to

Table 6.1 continued

work quickly and was generally reluctant to review what he had completed. (This was also apparent during the STEP tests discussed below.) His Standard Score of 11 on the subtest Coding was also *relatively* weak. This subtest requires students to copy symbols while being timed. Although the relatively lower score is not believed to be of concern, one could predict that Mark might not like to do paper-pencil types of tasks such as writing or working out problems. This prediction was verified when observing Mark during the STEP tests (see below). There is also a possibility that Mark might be relatively weak in attending to details and nonverbal cues, especially when compared to his verbal precocity. Mrs. M. confirmed that this observation was accurate.

Mark demonstrated superior skills, excellent concentration, little impulsivity, and a desire to do well on the Verbal tasks. The *relatively* weaker Performance subtest scores reflect some impulsivity, and a reluctance to check over his final product; such behaviors are often demonstrated by children who are not sufficiently challenged by the material.

On the Abstract Reasoning subtest of the *Differential Aptitude Tests*, Mark correctly answered 41 out of 50 items. This score is at the 90th percentile when compared with Grade 8 boys (grade 8 is the youngest group for which norms are available).

Mark took the Intermediate Level of the *School and College Ability Test* (SCAT), a test of basic verbal and quantitative abilities. The Intermediate Level of the SCAT was developed for Grade 6 Spring through Grade 9 Fall. Mark's Verbal Reasoning Score was at the 85th percentile, and his Quantitative Reasoning Score was at the 99th percentile.

Mark's ability to succeed in an academic setting is clearly Very Superior when compared to other students his age. The tests of ability and aptitude are useful in defining his level of performance when compared to others, but they are not designed to indicate which specific areas of study Mark should pursue. To attain this end, we proceeded with a Diagnostic Testing → Prescriptive Instruction (DT→PI) model.

Following the diagnostic component of the DT→PI, appropriate levels of the *Sequential Tests of Educational Progress* (STEP) Basic Concepts and computation were administered. Students are given 40 minutes to complete each of these tests. In the DT→PI Model, students who score above the 85th percentile for the appropriate out-of-level norms are considered to have mastered the material.

Mark completed Level IV (Form A) of the *STEP Basic Concepts* test in 15 minutes, and answered all 50 items correctly. This score is at the 99th percentile when compared to second-semester fifth graders. He completed Level IV (Form A) of the *Computation* test in 20 minutes, and made two errors (earning a score at the 99th percentile when compared to second semester fifth graders). When asked to rework the problems, Mark quickly recognized his errors and reworked the problems correctly. The pattern of making more errors on the *Computation* test as compared to the *Basic Concepts* test is not unusual; often talented children are bored by computation tasks,

Assessment

Table 6.1 continued

---

and prefer to work problems in their heads or to guess at an answer. This pattern is also reflected in Mark's grades in school (A+ in Math Concepts, A- in Math Computation).

Since the Level 4 STEP tests were too easy for Mark, the next difficulty level, Level 3, was administered. On the *Basic Concepts* test Mark correctly answered 41 out of 50 items (90th percentile when compared to second-semester eighth graders); when asked to retry the nine items that he missed, he correctly answered six of them. The three items that he missed twice were in the areas of probability, algebraic relations, and geometry/logic. On the *Computation* test for Level 3 Mark correctly answered 50 out of 60 items; this score is at the 83rd percentile when compared to second semester eighth graders. On the second try, Mark correctly answered three of the ten items. The problems that Mark missed twice were given to Mark's parents who agreed to work on these topic areas with Mark.

CONCLUSION AND RECOMMENDATIONS:

Mark is a friendly, nine-year-old boy who seems to have many varied interests and enjoys being challenged academically. As evidenced by his scores on the RPM, WISC-R, DAT, and SCAT, Mark has Very Superior intellectual ability. Based upon these and the STEP results, we have generated the following recommendations:

1. Mark would benefit from a DT→PI approach to studying mathematics, preferably in cooperation with the school system where he is enrolled.

2. As discussed with Mark's parents, there are many individuals within a community who might be appropriate mentors for Mark. For example, an undergraduate or graduate student in mathematics would have the background for challenging Mark. Also, an engineer or professor of mathematics might be an appropriate mentor. Another possibility for a mentor is an especially talented high school teacher who has an *excellent* background in mathematics. Whoever agrees to serve as Mark's mentor should be in agreement with the DT→PI approach and thoroughly understand its advantages for students such as Mark. An important factor to consider is that the mentor and Mark be committed to a consistent schedule. We recommend weekly meetings throughout the school year.

3. We recommend that students work once a week with the mentor for a period of two or three hours either during the regular school day or outside of school. While other students are in math class, we suggest that Mark have the opportunity to go to the library, work on the computer, work the math homework assigned by his mentor, etc. This means that Mark would be removed from the regular math class and would not be required to complete the assignments or tests given in the regular math class. Of course, when enrichment activities such as classroom math games or computer programming are planned, Mark should be included. We recommend that Mark work on mentor-assigned homework while the other students are working on their math homework. Communication between parents, mentor, and classroom teacher will ensure that Mark receive credit for work completed. By using standardized subject-matter tests, the mentor can demonstrate to school personnel that Mark has mastered specific work.

Table 6.1 continued

4. We recommend that the purpose of testing be explained to Mark. That is, sometimes he will be tested to see how much he has learned, but other times he will be tested to see what still needs to be learned. These two reasons for testing reinforce each other. Like many extremely able students, Mark has developed a habit of working problems quickly in his head in order to stimulate and challenge himself, since the materials themselves do not challenge him. Mark will need reminders and reinforcements to check his work, pay attention to details, and write out his thought process when solving more difficult problems. Writing out solutions to problems will help him develop the good habits he will need when he studies higher level mathematics such as calculus.

5. We recommend that Mark continue to participate in contests such as the Mathematical Olympiads for Elementary Schools (MOES). This gives him the opportunity to interact mathematically and socially with other talented students. In a couple of years, Mark would benefit from other programs for talented youth such as the summer classes offered by the school system or university-associated programs such as those provided by local, state, and regional talent searches.

6. With Mark's parents, we discussed the benefits for the student of a cooperative relationship between school and parents. We recommend that Mark's parents continue to help Mark develop a positive identity with his school.

7. We recommend that Mark's parents contact us in one month to let us know what decisions have been made about Mark's education. We also recommend that Mark participate in a Talent Search when he is a seventh grader by taking the College Board Scholastic Aptitude Test (SAT). By earning a high score on this out-of-level test, he should qualify for a participation in an academic summer program.

Assessment

8 LONG-TERM PLANNING

Many parents and educators who contact us concerning mathematically talented 4- to 11-year-olds tell us that their children are beginning to dabble in algebra, set theory, or number theory. Although studying these advanced topics may be appropriate for young children who are ready, we generally urge them to go more slowly, making sure that the child has a solid foundation in elementary mathematics, and to consider the points discussed in this chapter.

Elementary students who are extremely talented in mathematics need time to develop the necessary cognitive structures that characterize mathematical maturity.

Many of the parents who have contacted us have described their children as being able to solve "algebraic" problems without having been formally introduced to algebra, specifically the concept of "unknowns." Often the first question these parents ask is, "How do I know when my child is ready to study algebra?" Although a child might be extremely talented, an elementary student is unlikely to have a thorough background in general mathematics, the structure of the number system, or arithmetical problem solving. The educational goal is "mathematical maturity" at the elementary level before moving on to junior high and high school mathematics such as algebra. Mathematical maturity is characterized by Piagetian formal operational thinking (Piaget & Inhelder, 1969).

Extremely few elementary students have the necessary cognitive structures needed for abstract mathematics. High school mathematics such as algebra, geometry, and calculus require high

Planning

level thinking skills rarely found in young students. Although extremely bright youngsters who are studying high school level mathematics may have the computational skills that allow them to solve complex problems, they may not completely understand the underlying concepts. For the less mathematically mature student who studies high level math too early, the mathematics becomes so abstract that he or she does not have the advanced cognitive skills to fully integrate the concepts. Such students are at risk for becoming frustrated and turned off to mathematics.

In order to start at the right point and ensure that students develop strong mathematical underpinnings, it is extremely important to assess carefully what skills and content have been mastered.

Pretesting is one important and effective mechanism for determining the correct starting place based upon what has already been learned. Extremely talented students can demonstrate mastery of skills that have not been formally taught to them; this is especially true for those extremely talented in mathematics. For such students, instruction should begin at the point where they have mastered the content and are ready for new material. The DT→PI model, described in Chapter 4 (see also Stanley, 1978, 1979a), provides a mechanism for attaining the necessary background while efficiently and consistently challenging talented youth, leading to the first course in high school algebra without undue haste.

Mathematically brilliant youths should study mathematics at their appropriate level of mental functioning and at a steady rate.

This does not mean racing through the standard sequence in truncated periods of time. There is no need to study mathematics intensely every day; one weekly two-hour session with a mentor may provide the challenges and stimulation that a student needs. Pacing of this sort helps avoid a situation in which a student will not have the opportunity to study mathematics for long periods of time. Some students may need to participate in after-school or summer programs in order to be adequately challenged. If a student is fortunate enough to have an excellent school mathematics program, however, it probably won't be necessary for him or her to take additional mathematics classes in after-school or summer programs.

Mathematically talented students need to balance their accelerated study of mathematics with the study of other academic subjects and participation in extracurricular activities.

We think it is critical for all students to spend time doing activities in other subjects, as well as playing with friends and spending time alone. One area that elementary children might find challenging is languages. Other subjects that they might enjoy include astronomy, computer programming, statistics, history (including history of mathematics), and geology. Many museums, universities, and schools offer summer and after school programs in these subject areas. Other programs are available on such diverse topics as photography, drama, music, and robotics. Sports, dance, and gymnastics provide opportunities for social interaction and gross motor development.

Teachers, mentors, clubs, and competitions can enrich an accelerated mathematics curriculum for talented youths.

Skilled mathematics teachers can offer supplemental problems that are more advanced than typical students can handle. A mathematics mentor can enrich or supersede the youth's mathematics curriculum and provide suitable pacing. Chapter 5 describes materials that teachers and mentors might use.

Participation in clubs and contests offers students an enriching opportunity to develop their mathematical maturity and a chance to interact with other mathematically talented students. Mathematically talented youths should consider every opportunity to hone their talents in competitions, from participating in the Mathematical Olympiads for Elementary Schools (see Appendix B) and MathCounts in junior high school to striving to become a member of the United States team in the annual International Mathematical Olympiad. Contests also provide an opportunity to talented students to interact with their academic peers. One such competition is the American Regions Mathematics League (ARML), a two-day event including team and individual competition.

In addition to moving ahead in mathematics and other subjects, students can study and understand the material at a deeper level than is typical. One high school student who participated in the International Mathematical Olympiad said, "The whole thing has given me a much stronger feeling for math, . . . a very strong foundation [in] elementary math. Stronger in some ways, probably, than many mathematicians who didn't spend so much time in elementary math"

Planning

(Dauber, 1988, p. 10). The contests mentioned above are discussed in Appendix D.

Moving ahead extremely fast in the mathematical sequence is likely to catapult the elementary student beyond the offerings of the school system before high school graduation.

In a typical program, the progression is two years of pre-algebra, algebra I, geometry, algebra II, and precalculus. If a student waits until 9th grade to take algebra I, he or she will be unable to study calculus until entering college unless he or she takes two mathematics courses during the same year in high school. Unfortunately, the elementary student who completes pre-algebra prior to seventh or eighth grade may have to slow down too much at some phase, perhaps not even taking mathematics courses until at the right grade level to resume the sequence.

The challenge is to plan for the eventuality that the student will move beyond the course offerings of the school system. If there is an excellent college nearby where the student can readily take regular college courses part-time without jeopardizing his or her high school education, this will not be a problem. After completing at least two semesters of calculus, the conventional progression is linear algebra, differential equations, probability theory and statistics, and the various branches of "pure" mathematics such as analysis, higher algebra, mathematical logic, number theory, and topology.

Academic programs offer varied opportunities for able students to forge ahead in mathematics.

The truly mathematically talented youngster whose special abilities are recognized early should be made ready by age 12 or 13 to study algebra and subsequent courses quickly and well. This can be accomplished through continuation of the DT→PI model of instruction (see Chapter 4) or in one of the local or regional, residential summer programs conducted in various parts of the country. Some of these fast-paced mathematics programs are offered during the school year to local students.

A number of universities and colleges sponsor Talent Searches and academic programs based on a model developed by Julian C. Stanley at Johns Hopkins University during the 1970's. To participate in these programs, seventh and eighth graders take the American College Testing (ACT) College Entrance Examination or the College Board's Scholastic Aptitude Test (SAT). Those scoring high enough (i.e., about 430 on the SAT-Verbal section and/or 500 on the SAT-Mathematics section) are eligible to participate in the residential or commuter academic programs. The score of 500 on the SAT-M is the 49th percentile for college-bound male high school seniors and the 64th percentile for college-bound females. A student younger than 13 who scores 500 or above probably has the cognitive skills needed to master algebra and the courses that follow.

Universities sponsoring academic programs include: Arizona State, California State (Sacramento), Denver, Duke, Johns Hopkins, Iowa State, Minnesota, Northwestern, Purdue, The University of Iowa, University of North Texas, and Wisconsin (Eau Claire). Addresses for these programs are listed in Appendix D. Several of these universities, as well as colleges and state departments of public instruction, sponsor programs for younger students.

Students who participate in these academic programs need to make inquiries regarding school credit for the special classes as well as the availability and scheduling of classes to complement the courses completed. Thus, those students must be certain that it will be possible for them to continue with mathematics through high school-- in the school system, at a local college, or with a mentor--*before* taking one of the special mathematics courses. We also encourage students to supplement their mathematics education by taking courses in physics, computer science, chemistry, and biology.

Before moving on to college coursework, students should take the most challenging coursework available at the high-school level.

Students should have exhausted the advanced coursework offered in their high schools before taking college courses. Taking honors courses or Advanced Placement Program courses will help ensure that students have the proper preparation for college classes.

Advanced Placement Program (AP) classes, offered in many high schools, give talented students the opportunity to do college-level work before graduating from high school. Courses are available in many subjects, including Calculus, Computer Science, Physics, Biology, Chemistry, French, American History, and Psychology. AP examinations are usually taken after completing an AP course, although some students take the examination without having taken a formal course (often because the course is not offered at their school, or they cannot fit it into their schedules).

Earning 4s or 5s on a number of AP exams before entering college will allow talented youths to shorten the number of undergraduate years, thereby saving money and permitting them to enter graduate school early. Others may choose to get a more varied, higher level undergraduate education in four years by sampling many courses, earning two or more majors rather than one, or completing a master's and bachelor's degree concurrently.

CONCLUSION

The goals for these students are proper pacing, proper sequencing, plenty of stimulation, time for planning and contemplation, appropriately planned challenges, and continual reinforcement for worthwhile achievements. These goals can be accomplished best without unseemly haste. Don't plunge the quite-young student precipitously into algebra, set theory, number theory, or the like. Let those subjects come in the natural sequence as his or her talents unfold. Take the long view that leads to steadily increasing achievement and deep intellectual satisfaction.

Note: We are indebted to Julian C. Stanley for providing the framework for this chapter. An earlier version of this chapter was published as "Acceleration and Enrichment for Mathematically Talented Youth: Eight Considerations," by Stanley, Lupkowski, and Assouline in *Gifted Child Today*, 1990, volume 13, pages 2-4.

Appendix A

Forms to Aid Educators and Parents

The forms in this appendix are useful for educators and parents in implementing mathematics programs for their talented students.

Form 1: Principal/Teacher Information Form

Form 2: Student Profile--Mathematics Form

Appendix A

PRINCIPAL/TEACHER INFORMATION FORM

TO: (name of student)'s classroom teacher and principal

FROM: Ann E. Lupkowski, Director, Julian C. Stanley Mentor Program

I am pleased to inform you that your student is eligible to participate in the Julian C. Stanley Mentor Program. The purpose of this memo is to inform you of the goals of the program and to ask your cooperation in helping it run smoothly.

Students are selected for the Mentor Program on the basis of parent or teacher recommendations, high aptitude and achievement in mathematics, and high motivation for studying mathematics. They are matched with mentors who are graduate students in mathematics at the University of North Texas. Groups of one to three students and their mentors meet up to two hours once a week in the evening and on weekends. Mentors assign homework that the students are required to complete before their next meeting.

The Mentor Program can be considered accelerative because students typically work at a level well beyond their age-mates. The goals of the program are to discover what topics in mathematics the students have already learned well, find where the gaps are in their knowledge of mathematics, and challenge students by providing more advanced material. Through a diagnostic testing/prescriptive instruction process, mentors present students with material at an appropriate level and pace for their advanced abilities in mathematics.

To facilitate students' participation in this selective program, I ask classroom teachers and principals to permit the following:

1. Students should be excused from regular classroom mathematics activities. Through the use of standardized tests, they have demonstrated that they have learned the material at their grade level. Therefore, additional drill and practice of that material is not needed. After students finish studying a topic, they will be required to demonstrate mastery on a standardized test before moving on to the next level. I am happy to provide copies of mathematics achievement testing reports to school personnel if parents give their consent. Student progress is also assessed continuously by mentors via homework and mentor-made tests.

2. While students in the regular mathematics class are working on their teacher-assigned homework, students in the Mentor Program should be permitted to work on their mentor-assigned homework.

Appendix A

3. When students in the regular class are participating in enrichment activities such as working on the computer or playing mathematics games, Mentor Program students should be encouraged to participate.

4. The issue of assigning grades needs to be resolved by local school personnel and parents.

5. Related to the issue of grades is the question of placement; if students participate in the Mentor Program this year, in what mathematics class will they be placed next year? Both of these issues need to be considered by all of the parties involved.

For more specific information about the program and mathematically talented elementary school students, see the articles "Applying a Mentor Model for Young Mathematically Talented Students" (Lupkowski, Assouline, & Stanley, 1990), and "Eight Considerations for Mathematically Talented Youth" (Stanley, Lupkowski, & Assouline, 1990).

Please contact me if you have any questions or if I can help in any way. I welcome your suggestions for improving the program and for facilitating communication among teachers, principals, other school personnel, parents, students, and mentors. Thank you.

Please check the appropriate category and sign below:

___ We have read the above information and are pleased to facilitate our student's participation in this selective program. We will permit him/her to work on mentor-assigned homework during the regular mathematics class.

___ We have read the above information and have more questions before we encourage our student to participate in this program.

Teacher   *Date*   *Principal*   *Date*

Appendix A

Student's Name: _____
Date: _____ Grade: ____

Student Profile--Mathematics Form

1.	Grasp of basic math skills in the curriculum	Repetition required				No repetition needed	
		0	1	2	3	4	5
2.	Completion of math curriculum assignments	Never				Always	
		0	1	2	3	4	5
3.	Knowledgeable about math concepts	Somewhat				More than other students	
		0	1	2	3	4	5
4.	Time spent in a math enrichment activity	Fulfills requirements				Goes beyond expectations	
		0	1	2	3	4	5
5.	Self-motivated in Math Enrichment	Never				Frequently	
		0	1	2	3	4	5
6.	Expresses own ideas in Math Enrichment Activities	Never				Frequently	
		0	1	2	3	4	5
7.	Likes working with others in Math Enrichment ideas	Sometimes				Almost Always	
		0	1	2	3	4	5
8.	Prefers to work independently in Math Enrichment Activities	Sometimes				Almost Always	
		0	1	2	3	4	5

9. Participates in Math
   Enrichment Activities: Active   Passive   Distractive

10. Uses a computer at home: Yes   No   If yes, how?
    (student response to teacher question)

Teacher Comments:

© Developed by Jean Kratz, Resource Specialist, The Connie Belin National Center for Gifted Education.

Appendix B

Mathematical Olympiads for Elementary Schools

Reprinted with permission from Dr. George Lenchner, Executive Director, Mathematical Olympiads for Elementary Schools.

For more Olympiad problems, contact
Book Order Department, Math Olympiads
125 Merle Avenue, Oceanside, NY 11572

Appendix B

MATHEMATICAL OLYMPIADS FOR ELEMENTARY SCHOOLS
1989 - 1990
PROBLEMS, ANSWERS, AND SOLUTIONS

OLYMPIAD 1

1. Time: 5 minutes

    How many x's are in the diagram at the right?

    ```
              x x x
            x x x x x
          x x x x x x x
        x x x x x x x x x
        x x x           x x x
        x x x x x     x x x x x
        x x x x x x x x x x x x x
        x x x x x x x x x x x x x x x
    ```

2. Time: 5 minutes
    A jar contains a large number of pennies. The pennies can be divided into equal shares among 3, 4, 5, 6, 7, or 8 children with no pennies left over each time. What is the least number of pennies the jar could contain?

3. Time: 5 minutes
    If a natural number is multiplied by itself, the result is called a perfect square. Thus, 1, 4, 9, 16, 25, 36, 49, ... are perfect squares and also consecutive because they follow in order. The number 1000 is between two consecutive perfect squares. Which one of these two squares is closer to 1000?

4. Time: 6 minutes
    I spent 2/3 of my money in store A. I then spent 1/3 of what remained in store B. When I left store B, I had $4. How much money did I have when I entered store A?

5. Time: 6 minutes
    If 6 is placed at the right end of a two-digit number, the value of the three-digit number thus formed is 294 more than the original two-digit number. What is the original two-digit number? (Hint: let the original two-digit number be AB)

OLYMPIAD 2

1. Time: 5 minutes
    When 24 is added to a number, the result is the same as when the number is multiplied by 3. What is the number?

Appendix B

2. Time: 4 minutes
Suppose all counting numbers are arranged in columns as shown at the right. Under what letter will the number 300 appear?

A	B	C	D	E	F	G
1	2	3	4			
	7	6	5			
8	9	10	11			
	14	13	12			
15	16	...				

3. Time: 5 minutes
Mrs. Bailey has equal numbers of nickels and quarters. If the value of the quarters is $1.80 more than the value of the nickels, what is the total value of all coins together in dollars and cents?

4. Time: 6 minutes
In the addition problem at the right, AB and BA each represent a two-digit number. If A and B stand for different digits, find A. (Be sure to give only the value of A in your answer.)

```
  B A
  A B
+ A B
  C A A
```

5. Time: 5 minutes
If all the odd numbers from 1 through 301 inclusive are written, how many times will the digit 3 appear?

OLYMPIAD 3

1. Time: 5 minutes
A container has 10 red disks, 10 white disks, and 10 blue disks, all of the same size. If I am blindfolded when I pick disks from the container, what is the least number of disks I must pick in order to be absolutely certain that there are three disks of the same color among those I have picked?

2. Time: 5 minutes
In the target at the right, ring A, ring B, and circle C have different point values. The sum of the point values of A and B is 23, of B and C is 33, and of A and C is 30. What is the sum of the point values of A, B, and C?

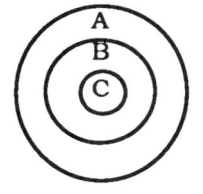

3. Time: 5 minutes
T and V in the four-digit number T37V represent different digits. If T37V is divisible by 88 without remainder, what digit is represented by T? (Hint: find the value of V first).

Appendix B

4. Time: 6 minutes
ABCD is a rectangle whose sides are 3 units and 2 units long. The length of the shortest path from A to C following the lines of the diagram is 5 units. How many different shortest paths are there from A to C?

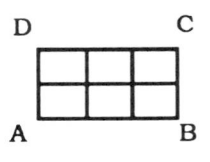

5. Time: 6 minutes
There are twice as many liters of water in one container as in another. If 8 liters of water are removed from each of the two containers, there will be three times as many liters of water in one container as in the other. How many liters of water did the smaller container have to begin with?

OLYMPIAD 4

1. Time: 5 minutes
The quotient of two number is 4 and their difference is 39. What is the smaller number of the two?

2. Time: 3 minutes
Express the fraction at the right as a simple fraction in lowest terms.

$$\cfrac{1}{3 + \cfrac{1}{3 + \cfrac{1}{3}}}$$

3. Time: 5 minutes
The structure at the right is made of unit cubes piled on top of each other. Some cubes are not visible. What is the number of cubes in the structure?

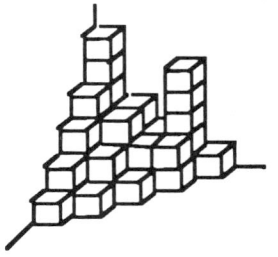

4. Time: 5 minutes
546 is one of six three-digit numbers each of which is different and has the digits 4, 5, and 6. What is the sum of these six numbers?

5. Time: 6 minutes
ABCD is a square with diagonal AC 8 units long. How many square units are in the area of the square?

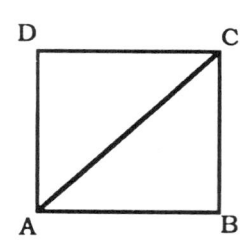

Appendix B

OLYMPIAD 5

1. Time: 5 minutes
   In the multiplication problem at the right, A, B, C, and D represent different digits, ABC is a three-digit number, and * means multiplication. What number is ABC? (Give your answer as a three-digit number.)

   ```
     A B C
   *     D
   -------
   1 2 3 4
   ```

2. Time: 5 minutes
   How many different rectangles can be traced using the lines in the figure given at the right?

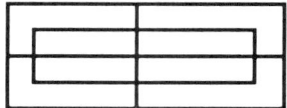

3. Time: 5 minutes
   Two students are needed to work in the school store during the lunch hour every day, and four students volunteer for this work. What is the greatest number of days that can be arranged in which no pair of the four students works together more than once?

4. Time: 5 minutes
   2, 3, 5, 7, 11, ... are examples of prime numbers each number has only itself and 1 as factors. Suppose the number of units in each of the length and the width of a rectangle are prime numbers and the perimeter is 36 units. What is the largest number of square units the area could have?

5. Time: 6 minutes
   When Paul crossed the finish line of a 60 meter race, he was ahead of Robert by 10 meters and ahead of Sam by 20 meters. If Robert and Sam continue to race to the finish line without changing their rates of speed, by how many meters will Robert beat Sam?

OLYMPIAD 1 ANSWERS AND SOLUTIONS

Answers: 1) 71   2) 840   3) 32 x 32 or 1024 or 32^2 4) $18 or 18   5) 32

Solutions:

1. Method 1
   Count the number of x's in each row:
   3 + 5 + 7 + 9 + 6 + 10 + 14 + 17 = 71

143

Appendix B

Method 2
(good for class discussion) Suppose the missing x's in the "triangular" region were filled in. Then there is a total of: 3+ 5 + 7 + 9 + 11 + 13 + 15 + 17 = 80 x's. (Notice that the average of the 8 addends is 10; or notice that the addends can be paired 3 and 17, 5 and 15, 7 and 13, 9 and 11, and that the sum of each pair is 20.) Since the triangular region was filled with 5 + 3 + 1 = 9 x's, then there are 80 - 9 = 71 x's in the diagram.

2. Find the least common multiple (LCM) of the numbers. Factor each of the numbers into primes; 2, 3, 2^2, 5, 2x3, 7, 2^3. The LCM of these numbers is 2^3 x 3 x 5 x 7 = 840.

3. The two squares are between 30 x 30 = 900 and 40 x 40 = 1600. Since 1000 is closer to 900, try 31 x 31 = 961, 32 x 32 = 1024. 31 x 31 and 32 x 32 are consecutive perfect squares and 1000 is between them. Clearly, 1024 is closer to 1000 than 961.

4. **Method 1**
Work backwards. The $4 that remained when I left store B was 2/3 of what I had when I entered store B. Since 2/3 of what I had was $4, then 1/3 of what I had was $2 and 3/3 of what I had was $6. This $6 is 1/3 of what I had when I entered store A. Then I had $18 when I entered store A.

Method 2
I had 1/3 of my money when I left store A. I spent 1/3 of that or 1/3 of 1/3 = 1/9 of my money in store B. I spent 2/3 in store A and 1/9 in store B, or a total of 2/3 + 1/9 = 6/9 + 1/9 = 7/9 of my money. I had 2/9 of my money or $4 left over. Then 1/9 of my money was $2 and 9/9 or all of my money was 9 x $2 or $18. I had $18 when I entered store A.

Method 3
Use a diagram to represent how much I had and shade the parts of the diagram that represented how much was spent in each store. The clear part of the diagram represents how much remained when I left store B. See last three sentences of Method 2 for rest of solution.

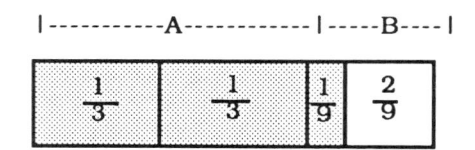

```
|-----------A------------|-----B----|
```

$\frac{1}{3}$	$\frac{1}{3}$	$\frac{1}{9}$	$\frac{2}{9}$

5. Let AB represent the original two-digit number. Then AB6 is the three-digit number which results when 6 is placed at the right end of AB. It is given that AB6 is 294 more than AB (see Method 1). This can also be expressed as the difference of AB6 and AB being 294 (see Method 2).

Appendix B

<u>Method 1</u>
In the units column, it is clear
that B = 2. It then follows that A = 3.

```
  A B
+ 2 9 4
-------
  A B 6
```

<u>Method 2</u>
Same as above.

```
  A B 6
-   A B
-------
  2 9 4
```

OLYMPIAD 2 ANSWERS AND SOLUTIONS

Answers: 1) 12   2) d   3) 2.70 or 270 cents   4) 4   5) 46

Solutions:
1. <u>Method 1</u>
Since the sum of the number and 24 is equal to 3 times the number, 24 must be equal to twice the number. The number must be 12.

<u>Method 2</u>
Represent the given information in a diagram. Clearly, 2N is 24. Then N = 12

N	24

N	N	N

<u>Method 3</u>
Algebra: Let N represent the number.
Given: (1) N + 24 = 3N
Subtract N from both members: (2) 24 = 2N
Divide both members by 2:   (3) 12 = N
Answer: then number is 12.

2. If each of the numbers shown is divided by 7, the numbers which have a remainder of 1 appear in column A, a remainder of 2 in column C, a remainder of 3 in column E, and so forth. Divide 300 by 7:

$$7 \overline{)300} = 42 \text{ R} 6$$

The remainder is 6; 300 will appear in column D.

3. <u>Method 1</u>
A quarter and a nickel differ in value by 20 cents, 2 quarters and 2 nickels differ in value by 2 x 20 cents or 40 cents, 3 quarters and 3 nickels differ in value by 3 x 20 cents or 60 cents, and so forth. Since the total difference of 180 cents equals 9 x 20 cents there must be 9 quarters and 9 nickels. Their total value is 9 x 30 cents or $2.70.

Appendix B

Method 2
Make a table of equal numbers of quarters and nickels and the difference in values. Let Q and N denote the number of quarters and nickels respectively, and D the difference in values.

Q	N	D
1	1	20c or 1x20c
2	2	40c or 2x20c
3	3	60c or 3x20c
.	.	...
?	?	180c or 9x20c

In the table we observe that the number of quarters or the number of nickels is the same as the number multiplying 20 cents in the D column. Since 180 cents = 9 x 20 cents, there must be 9 quarters and 9 nickels which have a total value of $2.70.

4. From the units column, B = 5. Then there is a carry of 1 in the tens column, and 1 + 5 + A + A must equal 10 + A, or 1 + 5 + A must equal 10. A must be 4.

5. Count the number of 3's that appear in the units place, then in the tens place, and then in the hundreds place.
Units place. In every ten consecutive numbers (1-10, 11-20, 21-30, ...), 3 appears once: 3, 13, 23, 33, 43, ... There are 30 groups of 10 consecutive numbers in 301. Therefore 3 will appear 30 times in the units place and each number is odd.
Tens place. In each hundred consecutive numbers (1-100, 101-200, 201-300) 3 appears in the tens place 10 times, or a total of thirty times:

30, 31, 32, 33, ..., 39,
130, 131, 132, 133, ..., 139,
230, 231, 232, 233, ...239,

But half of these numbers are even and half are odd. Then 3 appears in the tens place of the odd numbers 15 times.
Hundreds place. 301 is the only odd three-digit number in those being considered. Therefore 3 appears just one time in the hundreds place.
Answer: 3 appears as a digit 46 times in the set of odd numbers from 1 through 301 inclusive.

Appendix B

OLYMPIAD 3 ANSWERS AND SOLUTIONS

Answers: 1) 7   2) 43   3) 2   4) 10   5) 16

Solutions:
1. If I pick 6 disks blindfolded, there could be 3 disks of the same color among them. However, it is possible that I picked 2 disks of each color. Therefore I cannot be absolutely certain there are 3 disks of the same color among the 6 disks I picked. If I now pick a 7th disk, it must result in having 3 disks of the same color.

2. Method 1
Let a, b, and c denote the point values of A, B, and C, respectively. Then we need to find $a + b + c$.
Given: 1) $a + b = 23$
Given: 2) $b + c = 33$
Given: 3) $a + c = 30$
Add 1), 2), and 3):   4) $2a + 2b + 2c = 86$
Divide both members of 4) by 2:   5) $a + b + c = 43$

Method 2
Since $a + b = 23$ and $b + c = 33$, it follows that c is 10 more than a. Since $a + c = 30$, $a = 10$, $c = 20$. Then from either of the first two conditions, $b = 13$.
$$a + b + c = 10 + 13 + 20 = 43$$

3. Since the four-digit number is divisible by 88, it is also divisible by 8 and by 11. If a number is divisible by 8, the number formed by the last 3 digits of T37V (in this case 37V) is divisible by 8. Divide 37V by 8:

```
        4
    8 / 37V
        32
         5V
```

V must be 6. We know that the number T376 is also divisible by 11. We could try different values for T but it is easier to use the theorem for divisibility by 11: if a number is divisible by 11, the sum of the digits in odd-places (T + 7) and the sum of the digits in the even-places (3 + 6) must be equal or differ by a multiple of 11. $T + 7 = 3 + 6 = 9$. Therefore T must be 2.

Appendix B

4. Make a diagram for each path. It will be easier if you use a rule like "make the path go to the right as far as possible without repeating a previous path."

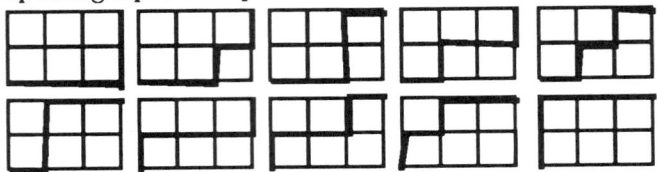

Answer: 10

5. Method 1
   Work backward

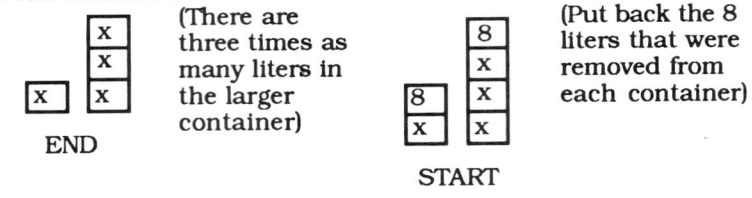

   |x|
   |x|
   |x|x|
   END

   (There are three times as many liters in the larger container)

   |8|
   |x|
   |8|x|
   |x|x|
   START

   (Put back the 8 liters that were removed from each container)

   Since there were twice as many liters in the larger container at the start, the condition is satisfied if x = 8 liters. So the containers had 16 liters and 32 liters to begin with
   Answer: the smaller container had 16 liters.

   Method 2
   Let the smaller container have x + 8 liters to begin with.

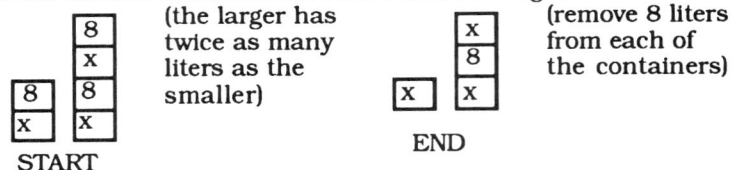

   |8|
   |x|
   |8|8|
   |x|x|
   START

   (the larger has twice as many liters as the smaller)

   |x|
   |8|
   |x|x|
   END

   (remove 8 liters from each of the containers)

   Since the larger had three times as many at the end, x must be 8. The container had 16 and 32 liters of water to begin with; and the smaller had 16 liters of water.

   OLYMPIAD 4 ANSWERS AND SOLUTIONS

   Answers: 1) 13 2) 10/33 3) 38 4) 3330 5) 32

   Solutions:
   1. Method 1
      If the quotient of the two numbers is 4, the larger must be 4 times the smaller, and the difference of the two numbers must be 3 times the smaller. Since the difference is 39, 3 times the smaller is 39. The smaller number must be 13. (The larger number is 4 times the smaller and must therefore be 52.)

Appendix B

Method 2
(Algebra) Let the smaller number be n. Then the larger number is 4n.

Given: 1) $4n - n = 39$
Simplify left side of 1):   2) $3n = 39$
Divide both members of 2) by 3:   3) $n = 13$
Answer: the smaller number is 13.

2. This is an "extended unit fraction." Work backward.

$$\cfrac{1}{3+\cfrac{1}{3+\cfrac{1}{3}}} = \cfrac{1}{3+\cfrac{1}{\frac{10}{3}}} = \cfrac{1}{3+\frac{3}{10}} = \cfrac{1}{\frac{33}{10}} = \frac{10}{33}$$

3. Method 1
Consider the structure as a collection of vertical columns of cubes. Count the number of columns having the same height and determine the number of cubes contained in theses columns. The tallest vertical column contains 6 cubes; the smallest vertical column has just 1 cube.

number of cubes in the vertical column	number of columns	total number of cubes
6	1	6
5	1	5
4	1	4
3	3	9
2	5	10
1	4	4
		total 38

Method 2
Slice" the structure mentally as shown at the right. Count the number of cubes in each slice.

slice	number of cubes
1	16
2	9
3	5
4	7
5	1
	total 38

Appendix B

4. List the six three-digit numbers each of which contains the digits 4, 5, and 6. Notice that each of the digits 4, 5, and 6 appears twice in each column. Therefore the sum of the digits in each column (without carrys) is 30. The sum is 3330.

```
 456
 465
 546
 564
 645
 654
3330
```

Method 1
Draw diagonal BD, four congruent right triangles are created, each with area (1/2) x4x4 = 8. Therefore the area of the square is 4 x 8 = 32 square units.

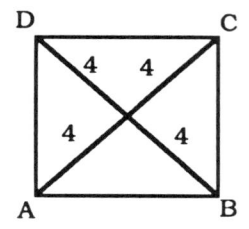

Method 2
Draw diagonal BD as above and regroup the 4 congruent squares as shown at the right to form 2 squares. Each of the 2 squares has area 16, or a total of 32.

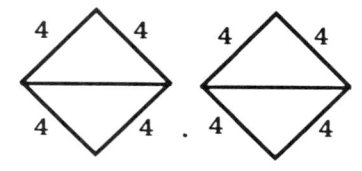

Method 3
Rearrange the 2 triangles given in square ABCD as shown at the right and add a dotted line to complete a square. The sum of the areas of the 2 given triangles is half of the area of the square obtained by adding the dotted line. The area of that square is 8 x 8 = 64. Then the sum of the areas of the 2 given triangles is 32.

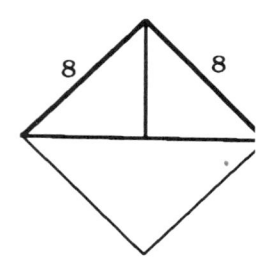

Appendix B

OLYMPIAD 5 ANSWERS AND SOLUTIONS

Answers: 1) 239   2) 18   3) 6   4) 77   5) 12

Solutions:
1. Strategy: find D first, and then find ABC by dividing 1673 by D. Possible values for D are 1, 2, 3, 4, 5, 6, 7, 8, or 9   Clearly D can't be 1. Otherwise the product would be the three-digit number ABC. D can't be 2, 4, 6, or 8. Otherwise, the product would be an even number. D can't be 5. Otherwise the product would have to end in 5 or 0. D can't be 3. Otherwise, the sum of the digits of the product 1673 would have to be a multiple of 3. Similarly, D can't be 9. Otherwise, the sum of the digits of the product would have to be a multiple of 9. The only digit that remains is 7. Then D = 7 and ABC = 1673/7 = 239.

2. Notice that the rectangle contains 8 separate regions: a,b,c, d,e,f,g,h. Some of these regions are rectangles. Other rectangles can be formed by combining two or more regions.

Regions which form a rectangle		Number of rectangles
1 region:	(a), (b), (c), (d)	4
2 regions:	(a,b), (b,c), (c,d), (d,a)	8
	(a,f), (b,g), (c,h), (d,e)	
4 regions:	(f,a,b,g), (e,d,c,h)	5
	(f,a,d,e,),(g,b,c,h)	
	(a,b,c,d)	
8 regions:	(a,b,c,d,e,f,g,h)	1
		total 18

3. The answer to this problem is equal to the number of different pairs that can be formed with the four students. Denote the different students by A, B, C, and D. Then the different pairs are:
> AB, AC, AD,
> BC, BD,
> CD

Since there are 6 different pairs, the n there are 6 days where it can be arranged that the same pair of students do not work together. (Note that AB and BA stand for the same pair; AB and BC are different pairs.)

151

Appendix B

4. The sum of the length and width (semiperimeter) is 18. We need to find two primes to represent the length and the width with sum 18.

Width	Length	Area
5	13	65
7	11	77

The largest area the rectangle could have is 77 square units.

5. When Paul finished the 60 meter race, Robert was at the 50-meter mark and Sam was at the 40-meter mark. This means that for every 5 meters that Robert runs, Sam runs 4 meters; or, for every 10 meters that Robert runs, Sam runs 8 meters. So, when Robert ran 10 meters to complete the race, Sam ran 8 meters to be at the 48-meter mark. Robert beats Sam by 12 meters.

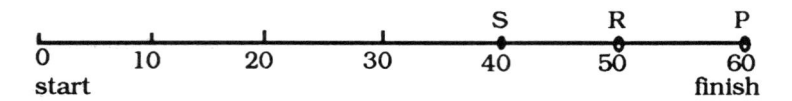

```
                                   S    R    P
|-------|-------|-------|-------|--•----•----•
0      10      20      30      40   50   60
start                                    finish
```

MATHEMATICAL OLYMPIADS FOR ELEMENTARY SCHOOLS
1990 - 1991
PROBLEMS AND ANSWERS

OLYMPIAD 1

1. Time: 3 minutes
ABCD represents a four-digit number. The product of its digits is 70. What is the largest four-digit number that ABCD can represent?

2. Time: 3 minutes
Express the value of the following expression in simplest form:

$$(5\tfrac{1}{3} - 2\tfrac{1}{2}) + (5\tfrac{1}{2} - 3\tfrac{1}{3})$$

3. Time: 5 minutes
At the right is a 3 by 3 by 3 cube. Suppose the entire outside of the cube is painted red. How many different 2 by 2 by 2 cubes with exactly three red faces can be found in the shown cube? (Note: Not all of the cubes are visible.)

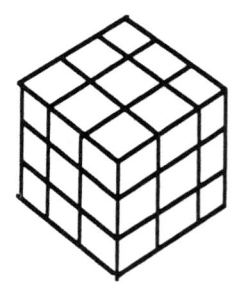

Appendix B

4. Time: 6 minutes
The cold water faucet of a bath tub can fill the tub in 15 minutes The drain, when opened, can empty the full tub in 20 minutes. If the tub is empty and the faucet and drain are both opened at the same time, how long will it take to fill the tub?

5. Time: 6 minutes
In the multiplication example at the right, different letters represent different digits and a blank space may represent any digit. The product WHAT is a four-digit number less than 5000. Find the number that WHAT represents.

```
        U T
        U T
      ─────
      _ _ 6
      _ 8
      ─────
      W H A T
```

OLYMPIAD 2

1. Time: 3 minutes
A stamp collector bought a rare stamp for $30, sold it for $42, bought it back for $50, and finally sold it for $48. Did the stamp collector make or lose money and how many dollars were made or lost?

2. Time: 4 minutes
The numbers M and N are different numbers selected from the first twenty-five counting numbers: 1, 2, 3, 4, ..., 24, 25. If M is larger than N, what is the smallest possible value that the expression at the right can have?

$$\frac{M \times N}{M - N}$$

3. Time 5 minutes
ABCD is a square which contains nine small congruent squares as shown. If the are of square ABCD is 36 units, what is the area of triangle ACE in square units?

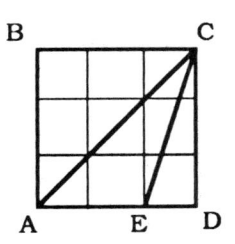

4. Time: 5 minutes
A prime number is a whole number, greater than 1, that is divisible only by itself and 1. Some examples of prime numbers are 2, 3, 5, 7, 11, and 13. What is the largest prime number, P , such that 9 times P is less than 400?

153

Appendix B

5. Time: 6 minutes
A person made a purchase for D dollars and C cents, and gave the cashier a $20 bill. The cashier incorrectly charged the person C dollars and D cents, and returned 4.88 in change. If the cashier had charged the correct price, what would the correct change have been?

OLYMPIAD 3

1. Time: 3 minutes
Suppose the counting numbers from 1 through 100 are written on paper. What is the total number of 3's and 8's that will appear on the paper?

2. Time: 3 minutes
Find the counting number that is equivalent to
1990 x 1991 - 1989 x 1990

3. Time: 5 minutes
Suppose 3! means 3 x 2 x 1, 4! means 4 x 3 x 2 x 1, 5! means 5 x 4 x 3 x 2 x 1, and so forth. Find the value of the following expression in simplest form:
$$\frac{8! - 6!}{3! \times 5!}$$

4. Time: 6 minutes
ABCD is a square; E, F, G, and H are midpoints of AP, BP, CP, and DP respectively. What fractional part of the area of square ABCD is the area of square EFGH?

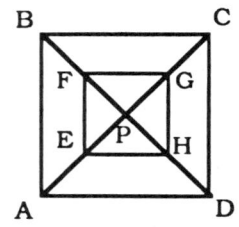

5. Time: 7 minutes
A bus was rented at a fixed cost by a group of 30 people. When 10 people were added to the group, the fixed cost of the bus did not change, but the charge for each person was $2 less than before. If each person paid the same charge as each of the others, find the cost of renting the bus in dollars.

OLYMPIAD 4

1. Time: 3 minutes
Suppose two days before yesterday was Wednesday. What day of the week will it be 100 days from today?

2. Time: 5 minutes
The first fifteen multiples of 6 are: 6, 12, 18, 24, ..., 84, 90.
What is the sum of these multiples of 6?

Appendix B

3. Time: 6 minutes
A, B, and C represent different digits, and a blank space may represent any digit in the division problem at the right. If C does not equal 0, what digits do A, B, and C each represent? (Indicate which digit goes with each letter.)

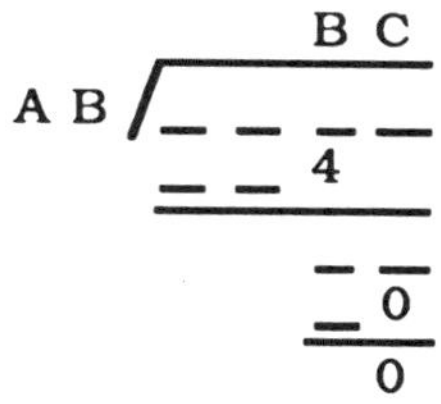

$$AB \overline{\smash{\big)}\begin{array}{r}BC\\ _____\end{array}}$$
$$\underline{\_\_\ 4}$$
$$\underline{\_\_\ 0}$$
$$0$$

4. Time: 6 minutes
Find the largest factor of 2520 that is not divisible by 6.

5. Time: 6 minutes
Five disks numbered 1, 2, 4, 8, and 16, are placed in a bag. Three disks are withdrawn from the bag, the sum of their numbers is recorded, and the three disks are then returned to the bag. If this process is repeated indefinitely, what is the largest number of different sums that can be recorded?

OLYMPIAD 5

1. Time: 4 minutes
Consecutive odd numbers are odd numbers that follow in order such as 5, 7, 9, and 11. If the sum of five consecutive odd numbers is 85, what is the largest number of the five?

2. Time: 5 minutes
The three-digit number 104 has a digit-sum of 1 + 0 + 4, or 5. How many different three-digit numbers, including 104, each have a digit-sum of 5?

3. Time: 6 minutes
A light flashes every 2 minutes, a second light flashes every 2.5 minutes, and a third light flashes every 3 minutes. If all three lights flash together at 9 a.m., what is the next time they will all flash together?

4. Time: 6 minutes
ABCD is a square with each side divided into three segments of length 1 unit, 8 units, and 1 unit respectively, as shown in the diagram. What is the sum of the areas of the four shaded triangles?

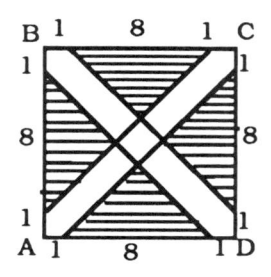

155

Appendix B

5. Time: 6 minutes
A fast clock gains one minute per hour and a slow clock loses 2 minutes per hour. At a certain time, both clocks are set to the correct time. Less than 24 hours later, the fast clock registers 9 o'clock at the same moment that the slow clock registers 8 o'clock. What is the correct time at this moment?

OLYMPIAD ANSWERS

Answers to OLYMPIAD 1:
    1) 7521
    2) 5
    3) 8
    4) 60 minutes or 1 hour
    5) 1296

Answers to OLYMPIAD 2:
    1) made 10 or $10
    2) 25/24 or 1 and 1/24
    3) 12
    4) 43
    5) $7.85 or 785 cents

Answers to OLYMPIAD 3:
    1) 40
    2) 3980
    3) 55
    4) 1/4
    5) $240

Answers to OLYMPIAD 4:
    1) Monday
    2) 720
    3) $A = 1, B = 8, C = 5$
    4) 315
    5) 10

Answers to OLYMPIAD 5:
    1) 21
    2) 15
    3) 9:30, or 9:30 a.m.
    4) 64
    5) 8:40, or 20 minutes before 9, or equivalent.

Appendix C

The Constructivist Approach to Educating Mathematically Talented Students

The constructivist perspective on learning mathematics is that knowledge is **constructed** by the child after reflection on the exploration of mental and physical actions. This is in contrast to a more traditional philosophy about learning, which assumes that knowledge is received from the environment. Key terminology in the constructivist approach is the **active involvement** of the learner.

Mathematics, as traditionally taught, has focused on a sequential presentation of pre-determined material. The constructivist approach is open-ended, and requires involvement of the student with his/her own thought processes as well as with those of other students. In contrast to the assumption that mathematics is a body of knowledge that is waiting to be received by the learner, the constructivist assumes that mathematics is a learning activity that requires that students talk about the activity with each other.

"Constructivist" teachers have as their goal the development of their students as autonomous creators of their own mathematics. In constructivism, children are regarded as responsible for their own learning, as it is their responsibility to create their own meanings and to learn to negotiate those meanings with their peers. The activity of mathematics is *constructing* relationships and *building* algorithms. If children cannot construct their own notions about mathematics, then they cannot be "taught" in the traditional sense of didactically

Appendix C

conveying information.

For more information on the constructivist approach to mathematics see:

Young Children Reinvent Arithmetic: Implications of Piaget's Theory, by Constance Kamii with G. DeClark (1985), New York: Teacher's College Press

Enquiring Teachers Enquiring Learners: A Constructivist Approach for Teaching, by C. Fosnot (1989), New York: Teacher's College Press

Teachers and Students: Constructivists Forging New Connections, by J. Brooks, in *Educational Leadership*, (February, 1990, pp. 68-71)

Constructivist Learning and Teaching, by D. Clements, and M. Battista, *Arithmetic Teacher*, (September, 1990, pp. 34-35)

Appendix D

Programs and Resources

In this appendix, we have classified and organized information and resources that are relevant to mathematically talented elementary students. We have included several annotated lists and suggestions for obtaining further information.

Appendix D

NATIONAL ORGANIZATIONS

Federal Office of Gifted & Talented Education, 555 New Jersey Ave. (Room 504), Washington, D.C., 20208. Phone: (202) 219-2169. The Federal Office disseminates information and administers grants to schools, universities, and other agencies.

Gifted Students Institute (P.O. Box 23029, Texas Woman's University, Denton, TX 76204; phone, (817)898-2210) holds an annual conference and publishes a magazine.

Mathematical Association of America (1529 Eighteenth St., NW, Washington, DC 20036; phone, (202)387-5200) sponsors American high school mathematics competitions leading to the International Mathematical Olympiad.

National Association for Gifted Children (NAGC), 1155 15th St., NW, Suite 1002, Washington, DC 20005. Phone: (202) 785-4268. NAGC is the largest national group organized on behalf of gifted and talented children. It sponsors conferences and workshops, disseminates information, and publishes materials of interest to parents and teachers of the gifted. Members receive a subscription to Gifted Child Quarterly and the membership newsletter. Most states have a division of NAGC.

National Council of Teachers of Mathematics (1906 Association Dr., Reston, VA 22091; phone (703)620-9840) publishes many useful materials, including the journals, Mathematics Teacher and Arithmetic Teacher.

The Association for the Gifted (TAG) is a division of the Council for Exceptional Children (1920 Association Drive, Reston, VA 22091; phone, (804)620-3660). TAG holds annual conference in conjunction with the national CEC meeting, and disseminates information about gifted and talented children. The Journal for the Education of the Gifted is the professional journal of TAG.

Appendix D

STATE AND LOCAL GROUPS

The gifted/talented teacher in your district can help you find information about state and local groups organized on behalf of gifted and talented children. Contact the National Association for Gifted Children or your state's Department of Education for the name and address of the director of gifted/talented education in your state. He or she can give you information on programs and resources available locally. If there is no organized group in your area, start one! Many state and local groups have a special division for parents of gifted students.

JOURNALS, MAGAZINES, AND NEWSLETTERS

Arithmetic Teacher (for elementary level) and *Mathematics Teacher* (for secondary level) are published by the National Council of Teachers of Mathematics. (1906 Association Drive, Reston, VA 22091)

Gifted Child Quarterly is published by the National Association for Gifted Children (NAGC) and provides research and theory, book reviews, and descriptions of educational practices. The address is 1155 15th St. NW, Suite 1002 Washington, D.C. 20005.

Gifted Child Today (P.O. Box 6448, Mobile, Alabame 36660-0448, is a magazine written especially for parents and teachers.

Gifted Children Monthly (Gifted and Talented Publications, Inc., 213 Hollydell Dr., Sewell, NJ 08080) is newsletter for parents.

Journal for the Education of the Gifted (JEG) is a scholarly journal reporting research and reviews of research. It is the official publication of The Association for the Gifted (TAG). The University of North Carolina Press, P.O. Box 2288, Chapel Hill, North Carolina 27515-2288.

Roeper Review (P.O. Box 329, Bloomfield Hills, MI 48303; provides scholarly reviews of research on gifted/talented children and reports of successful programs.

Appendix D

Understanding Our Gifted, (P.O. Box 18268, Boulder, CO 80308-8268) is a newsletter for parents, teachers, and other interested persons. Topics include parenting, instructional strategies, new developments in research, children's literature, and legislative updates.

TALENT SEARCHES AND SUMMER PROGRAMS

A number of universities and colleges sponsor talent searches and summer programs based on programs developed by Julian C. Stanley at Johns Hopkins University during the 1970's. Typically, seventh graders take the College Board's Scholastic Aptitude Test (SAT) in January. Those scoring high enough (usually at least 400 on the verbal section and 450 on the mathematical section) are eligible to participate in the residential or commuter academic programs. Many of the organizations listed below also sponsor programs for younger students. Check local university, museums, and schools for information about other programs.

Academic Talent Program, School of Education, California State University, 6000 J St., Sacramento, CA 95819.

Center for Academic Precocity, College of Education, Arizona State University, Tempe, AZ 85287-2711.

Center for the Advancement of Academically Talented Youth, Johns Hopkins University, Charles St., Baltimore, MD 21218.

Center for Talent Development, School of Education and Social Policy, Northwestern University, 2003 Sheridan Rd., Evanston, IL 60201.

Challenges for Youth--Talented and Gifted, N157 Lagomarcino, Iowa State University, Ames, IA 50011.

The Connie Belin National Center for Gifted Education, 210 Lindquist Center, University of Iowa, Iowa City, Iowa 52242.

Hampshire College Summer Studies in Mathematics, Hampshire College, Box NS, Amherst, MA 01002.

Appendix D

Mathematics Talent Development Project, University of Wisconsin at Eau Claire, 3216 S. Lexington Ave., Eau Claire, WI 54701.

Professor Arnold E. Ross's Mathematics and Science Program for High Ability Pre-College Students, Department of Mathematics, The Ohio State University, Columbus, OH 43210.

Research Science Institute, 7710 Old Springhouse Road, McLean, VA 22102.

Rocky Mountain Talent Search Summer Institute, Bureau of Educational Services, MRH 114, University of Denver, Denver, CO 80208.

Talent Identification Program, Box 4077, Duke University, Durham, NC 27706.

University of Minnesota Talented Youth Mathematics Program, UMTYMP, Special Projects, School of Mathematics, 115 Vincent Hall, 206 Church St. SE, University of Minnesota, Minneapolis, MN 55455.

MATHEMATICS CONTESTS

The mathematics contests below are listed in chronological order from elementary school through university levels:

MOES, Mathematical Olympiads for Elementary Schools, 125 Merle Ave., Oceanside, NY 11572.

MATHCOUNTS, 1420 King St., Alexandria, VA 22314.

AJHSME, American Junior High School Mathematics Exam, American Mathematics Competitions, 1714 Vine Street, University of Nebraska, Lincoln, Nebraska 68588.

ARML, American Regions Mathematics League, 711 Amsterdam Avenue, New York, NY 10025.

Appendix D

AHSME, American High Shcool Mathematics Exam, Mathematical Association of America, 1529 18th St., N.W., Washington, D.C. 20036.

INTERNATIONAL MATHEMATICAL OLYMPIAD, Mathematical Association of America, 1529 18th St., N.W., Washington, D.C. 20036.

WILLIAM LOWELL PUTNAM MATHEMATICAL COMPETITION, Mathematical Association of America, 1529 18th St., N.W., Washington, D.C. 20036.

OTHER CONTESTS

Invent, America! 510 King Street, Suite 420, Alexandria, Virginia 22314.

Science Olympiads for Grades K-12. A series of intramural, district, regional, state, and national science contests for three age divisions: grades K-5, 6-9, and 9-12. Contact Executive Secretary of the Science Olympiad, Inc., 5955 Little Pine Lane, Rochester, MI 48064 for membership materials and information.

International Chemistry Olympiad. High-level competition for high school students. Students nominated by their local selection chairperson take a national test; the top 20 scorers on this exam attend a 10-day study camp held at the U.S. Air Force Academy. Four students are chosen to represent the United States at the international competition. The ICO consists of theoretical problems and laboratory experiments. Gold, silver, and bronze medals are awarded to top scorers. Contact the American Chemical Society, Education Dept., 1155 16th St., N.W., Washington, D.C. 20036.

International Physics Olympiad. Competition for high school students. Qualifying exams are administered to students nominated by their physics teachers. The top 50 scorers on this test are given another exam. The 20 top scorers on the second exam are invited to attend a two-week training session held at the University of Maryland. Five finalists represent the United States at the international competition. The Olympiad consists of a two-part exam of theoretical

Appendix D

and experimental problems. American Association of Physics Teachers, 5112 Berwin Road, College Park, MD 20740.

Westinghouse Science Talent Search. High school seniors may submit a science or mathematics project by December 15th. College scholarships awarded range from $1,000 to $20,000. Science Service, 1719 N St., NW, Washington, DC 20036.

ADVANCED COURSES

Advanced Placement Program
AP exams, given each year in May, are a marvelous way for students to upgrade their curriculum, get a better all-round education, enter college ready for sophomore-level courses, and to save money on college costs. Students are not required to take a formal AP course in order to take the exam. Most of the top colleges in the country award college credit and advanced course placement for a score of 3, 4, or 5 on an AP exam. See Patricia Lund Casserly's What College Students Say About Advanced Placement, available from the Advanced Placement Program, P.O. Box 6670, Princeton, NJ 08541-6670.

College-Level Correspondence Courses
Although it is preferable for students to take courses in a "face-to-face" situation with a teacher or mentor, sometimes that is not possible. Correspondence courses offer an opportunity for highly motivated students to study more advanced material than is available in their home schools. Check with a state university in your region for more information.

BOOKS OF RESOURCES

Mountains to Climb. National Association for Gifted Children, 5100 N. Edgewood Drive, St. Paul, Minnesota 55112.

Directory of Science Training Programs for High Ability Precollege Students. Science Service, Inc., 1719 N St., NW, Washington, DC 20036.

Appendix D

For a list of resources that are especially helpful for high school students, write for the College Board Publications Catalog (The College Board, 45 Columbus Ave., New York, NY 10023-6992).

PUBLISHERS

Helpful materials can be obtained from the following publishers:

AMSCO School Publications, 5th Floor, 315 Hudson St., New York, NY 10013.

Creative Publications, 5040 West 111th St., Oak Lawn, IL 60453.

Dale Seymour Publications, 1100 Hamilton Court, Menlo Park, CA 94025.

D.O.K Publishers, P.O. Box 605, East Aurora, N.Y. 14052.

Free Spirit Publishing, 400 First Ave North, Suite 616, Minneapolis, MN 55401.

GCT Inc., 314-350 Weinacker Ave., PO Box 6448, Mobile, AL 36660-0448. Catalog lists materials for gifted, creative, and talented children and youth. Books, games, and other resources.

The Learning Company, 6493 Kaiser Dr., Fremont, CA 94555.

The Mentor Ship, 18515 West 66th Place, Shawnee, KS 66218.

Critical Thinking Press & Software, Box 448, Pacific Grove, CA 93950.

National Council of Teachers of Mathematics, 1906 Association Dr., Reston, VA 22091. Publishes books helpful in teaching mathematics.

Appendix D

Sunburst Communications, 39 Washington Ave., Pleasantville, NY 10570.

Trillium Press, First Ave, Unionwille, NY 10988.

Zephyr Press, 3865 E. 34th St., #101, P.O. Box 13448-A Dept. 11, Tucson, AZ 85732-3448.

Appendix D

Glossary

Above-Average: A "normal distribution" of students is shaped like a bell curve. If 100 student scores are plotted on a graph, a bell curve would likely be generated. Most of the scores earned by the students would fall in the middle of the curve. Average scores are those in the middle two-thirds of the curve. Students who earn scores **above** the average scores are **above-average**. Compared to the other 100 scores, their scores are better than 85 percent of the scores. In contrast, scores earned by "gifted" students are considered to be those in the top 5 percent. Compared to 100 students, gifted students earned scores that were better than 95 percent of those students.

Acceleration: Moving faster than is typical through the curriculum. This can be accomplished by entering kindergarten, first grade, or college early ("early admission"); skipping an entire grade ("grade-skipping"), or advancing in a subject area more rapidly than is typical ("subject-matter acceleration"). The benefits and drawbacks of accelerating students are considered in Chapter 3.

Achievement Test: Tests usually given to students after completing a course or topic. They are used to evaluate students' accomplishments. Results from achievement tests are also useful for adapting instruction to individual needs because examiners can determine what topics a student does or does not know.

Age-Equivalent: Children's scores on tests are sometimes compared to the performance of a typical child at a given age level. A child's score on a test will correspond to the highest year level that he or she can successfully complete. However, caution needs to be used in drawing

Glossary

conclusions regarding most age-equivalent scores, as scores at the extremes are usually extapolations, and not actually based upon performance. For example, a six-year-old who earns an age-equivalent of 14 years on a reading test has actually earned an age-equivalent score that is based upon an extrapolation of data, not upon an assessment of 14-year-olds.

Age-in-grade grouping: Placing a student with other students of the same age (i.e., birth year) and grade level.

Aptitude Tests: Used to predict performance in specific areas such as verbal comprehension, mathematical ability, mechanical aptitude, spatial relations aptitude, or nonverbal reasoning ability. Distinguishable from intelligence tests in that they have more specific content (i.e., they measure only one or a few abilities rather than a wide variety of abilities).

Beyond-Level Testing: Giving a test that was developed and normed for students who are several years older. Also called out-of-level testing or above-level testing.

Ceiling Effect: Occurs when a test does not have enough difficult items to measure a student's abilities accurately. Students who get all or almost all of the items right on a test are said to have reached the "ceiling" of the test.

Curriculum Compacting: A method developed for eliminating unnecessary repetition of material already learned. What a student already knows is determined and then the student is allowed to move on. The time saved can be used for engaging in enrichment activities or for studying more advanced material.

Curriculum-Based Instruction (CBI): Performance is based upon mastery of the curriculum rather than on a standardized comparison with other students (which is a normative comparison).

Diagnostic Testing -> Prescriptive Instruction (DT->PI) Model: A model for educating talented youth, that is, useful for determining specific strengths and weaknesses in a subject. This information is then related to instructional prescriptions. DT->PI model participants are first given aptitude tests. High-scoring students are recommended for further testing using beyond-level achievement tests. Instruction is prescribed based upon those test results. Students work on the topics they do not fully understand, and they spend little time on topics they already know well. After studying a topic, students are retested to assess their level of mastery. See Chapter 4 for a more complete discussion of this model.

Differentiated Curriculum: Curriculum that is designed specifically for gifted students and is defensibly different from the curriculum offered to average students. Can refer to a medley of educational programs including acceleration and enrichment.

Enrichment: Opportunities to study subjects or content not covered in the typical school curriculum and/or an in-depth exploration of topics covered in the regular curriculum. Students participating in enrichment activities study richer and more varied content.

Grade-Equivalent: The average score earned by children on a test is calculated. This information is used to develop grade norms. An individual child's score can be compared to that of the grade norms. A score of 5.0 refers to average performance compared to other students at the beginning of fifth grade. A score of 5.9 means a student's performance was comparable to that of the average student in the 9th month of the fifth grade year.

Intelligence Test: Designed to sample a wide variety of functions in order to estimate the individual's general intellectual level. See Chapter 7 for a more detailed discussion of various intelligence tests.

IQ (Intelligence Quotient): The score earned on an intelligence test. The average score on an IQ test is 100. Children with IQs of 130 and above are generally considered "gifted," and they constitute the upper 2 to 3 percent of the population.

Mentor: Mentors serve as instructors for talented students and are well-trained in the subject matter considerably beyond what they are expected to teach. Typically, mentors work with only one or two students at a time. They present new information to students rather than reviewing already-studied material the way a tutor would.

Glossary

Mentors help their students (sometimes called "mentees") to move through material at a challenging pace. Mentors not only present new information to students, they also clarify and extend the material to be learned. The goal of mentoring is to help students learn self-teaching skills, not to spoon feed them information. See Chapter 4 for more information about mentors.

Normed Sample (norm group): When tests are developed, they are given to a large group of students. This group comprises the norm group. Later, when individual students are tested, their performance can be compared to that of the first group of students, and the individual can be said to be average, above-average, or below-average when compared to the norm group.

Parallel Forms of a Test (alternate forms): Two or more tests containing different items that cover the same material at the same level of difficulty.

Percentile rank: Allows the student's score on a test to be compared to other children his or her age. An average score on a test yields a percentile ranking of 50, meaning that the child scored higher than 50 percent of the children in the comparison group. Talented children tend to score at the 97th percentile or above when compared to their own age group.

Precocious: Showing exceptionally early development. For example, a precocious mathematics student may multiply and divide at the age of four.

Prodigy: An extraordinarily talented youngster who performs adult-level accomplishments at a young age.

Psychological Assessment: Provides information about a student when compared to others of the same age. Psychological assessments can include intelligence testing, personality profiles, and information about social and emotional maturity. This topic is covered in more detail in Chapter 7.

Spiral Approach to Mathematics Curriculum: Students are presented with the material to be learned a portion at a time. Then, they come back to the same topic again and again, each time at a slightly more advanced level. For example, in first grade, students learn one-digit addition, in second grade they learn two-digit addition, and in third grade they learn three-digit addition. *We do not advocate this approach for talented youth because it is usually not challenging to bright students.*

Glossary

References

Adderholdt-Elliott, M. (1987). *Perfectionism: What's bad about being too good?* Minneapolis: Free Spirit.

Anastasi, A. (1979). Some reflections on the acceleration-enrichment controversy. In W. C. George, S. J. Cohn, & J. C. Stanley (Eds.), *Educating the gifted: Acceleration and enrichment.* Baltimore: Johns Hopkins University.

ASCD Update. (1990). *30*(9), p. 8.

Assouline, S. G., Colangelo, N., & Lupkowski, A. E. (1992). *The Iowa Acceleration Scale.* Iowa City: University of Iowa.

Assouline, S. G., & Lupkowski, A. E. (1992). Extending the Talent Search Model: The potential of the SSAT for identification of mathematically talented students. In N. Colangelo, S. G. Assouline, & D. L. Ambroson (Eds.), *Talent development: Proceedings from the 1991 Henry B. and Jocelyn Wallace National Research Symposium on Talent Development* (pp. 223-232). Unionville, NY: Trillium.

Baroody, A. J. (1987). *Children's mathematical thinking.* New York: Teachers College Press.

Bartkovich, K. G., & George, W. C. (1980). *Teaching the gifted and talented in the mathematics classroom.* Washington, D.C.: National Education Association.

References

Belcastro, F. P. (1990). Official position on vertical acceleration is inappropriate. *Roeper Review, 12*(4), 252-253.

Benbow, C. P. (1986). SMPY's model for teaching mathematically precocious students. In J. S. Renzulli (Ed.), *Systems and models in programs for the gifted and talented* (pp. 1-26). Mansfield Center, CT: Creative Learning Press.

Benbow, C. P. (1991). Mathematically talented children: Can acceleration meet their educational needs? In N. Colangelo & G. A. Davis (Eds.), *Handbook of gifted education* (pp. 154 - 165). Needham Heights, Mass: Allyn & Bacon.

Brody, L. E., Lupkowski, A. E., & Stanley, J. C. (1988). Early entrants to college: A study of academic and social adjustment during freshman year. *College and University, 63*, 347-359.

Carter, K. R., & Ormrod, J. E. (1982). Acquisition of formal operations by intellectually gifted children. *Gifted Child Quarterly, 26* (3), 110-115.

Clark, B. (1988). *Growing up gifted* (3rd ed.) Columbus: Ohio Psychology Publishing.

Clasen, D. R., & Hanson, M. (1987). Double mentoring: A process for facilitating mentorships for gifted students. *Roeper Review, 10*(20), 107-110.

Cognitive Abilities Test. (1987). Chicago: Riverside Publishing Company.

Cohn, S. J. (1988). Assessing the gifted child and adolescent. In C. Kestenbaum and D. Williams (Eds.), *Handbook of Clinical Assessment of Children and Adolescents*, Vol. 1 (pp. 355-376). New York: New York University Press.

Cohn, S. J. (1991). Talent searches. In N. Colangelo & G. A. Davis (Eds.), *Handbook of gifted education* (pp. 166 - 177). Needham Heights, Mass: Allyn & Bacon.

References

Colangelo, N. & Dettmann, D. F. (1981). A conceptual model of four types of parent-school interactions. *Journal for the Education of the Gifted.* 5(2), 120-126.

Colangelo, N. & Dettmann, D. F. (1985). Families of gifted children. In S. W. Ehly, J. C. Conoley, & D. Rosenthal (Eds.), *Working with parents of exceptional children* (pp. 233-255). St. Louis: Times Mirror/Mosby College Publishing.

Comprehensive Testing Program II technical information. (1983). Princeton, NJ: Educational Testing Service.

Cox, S. (1988). MATHCOUNTS: The ultimate in problem solving. *Arithmetic Teacher, 1,* 20-22.

Daniel, N., & Cox, J. (1988). *Flexible pacing for able learners.* Reston, VA: The Council for Exceptional Children.

Dauber, S. L. (1988). International Mathematical Olympiad. *Gifted Child Today.* September/October, 8-11.

Daurio, S. P. (1979). Educational enrichment versus acceleration: A review of the literature. In W. C. George, S. J. Cohn, and J. C. Stanley (Eds.), *Educating the gifted: Acceleration and enrichment.* Baltimore: Johns Hopkins University Press.

DeLeon, P. H., & VandenBos, G. R. (1985). Public policy and advocacy on behalf of the gifted and talented. In F. D. Horowitz & M O'Brien (Eds.), *The gifted and talented: Developmental perspectives.* Washington, D.C.: American Psychological Association.

Delisle, J. (1992). *Guiding the social and emotional development of gifted youth.* New York: Longman.

Delisle, J. & Galbraith, J. (1987). *The gifted kids survival guide II.* Minneapolis: Free Spirit.

Ehrlich, V. Z. (1985). *Gifted children: A guide for parents and teachers.* New York: Trillium Press.

References

Elkind, D. (1988). Acceleration. *Young Children, 43*(4), 2.

Feldman, D. H. (1982). A developmental framework for research with gifted children. In D. H. Feldman (Ed.), *Developmental approaches to giftedness and creativity: New directions for child development.* (pp. 31-45). San Francisco: Jossey-Bass.

Flanders, J. R. (1987). How much of the content in mathematics textbooks is new? *Arithmetic Teacher, 35*(1), 18-23.

Fox, L. H. (1974). Facilitating educational development of mathematically precocious youth. In J. C. Stanley, D. P. Keating, & L. H. Fox (Eds.), *Mathematical talent: Discovery, description, and development* (pp. 47-69). Baltimore: The Johns Hopkins University Press.

Gallagher, J. J., & Courtright, R. D. (1986). The educational definition of giftedness and its policy implications. In R. J. Sternberg and J. E. Davidson (Eds.) *Conceptions of giftedness* (pp. 93-111). Cambridge: Cambridge University Press.

Gardner, H. (1983). *Frames of mind.* New York: Basic Books.

Gross, M. (1986). Radical acceleration in Australia: Terence Tao. *G/C/T, 9*(4), 2-11.

Herman, K. E. (1982). Publicity and the prodigy. *G/C/T, 25,* 60-61.

Hildreth, G. (1966). *Introduction to the gifted.* New York: McGraw-Hill.

Hollingworth, L. (1942). *Children above 180 IQ Stanford Binet: Origin and development.* Yonkers-on-Hudson, NY: World Book.

Howley, A., Howley, C. B., & Pendarvis, E. D. (1986). *Teaching gifted children: Principles and strategies.* Boston: Little, Brown and Company.

House, P. A. (Ed.). (1987). *Providing opportunities for the mathematically gifted K-12.* Reston, VA: National Council of Teachers of Mathematics.

References

Hunt, J. M. (1961). *Intelligence and experience.* New York: Ronald Press.

Janos, P. M., & Robinson, N. M. (1985). Psychosocial development in intellectually gifted children. In F. D. Horowitz and M. O'Brien (Eds.) *The gifted and talented: Developmental perspectives.* Washington D.C.: American Psychological Association.

Kamii, C. (1985). *Young children re-invent arithmetic: Implications of Piaget's theory.* New York: Teachers College Press.

Kamii, C. (1989). *Young children continue to re-invent arithmetic.* New York: Teachers College Press.

Keating, D. P. (1975). Precocious cognitive development at the level of formal operations. *Child Development, 46,* 276-280.

Klausmeier, K. (1986) Enrichment: An Educational Imperative for Meeting the Needs of Gifted Students. In C.J. Maker (Ed.) *Critical issues in gifted education (V.1): Defensible programs for the gifted* (pp. 209-220). Austin: Pro-ed.

Krutetskii, V. A. (1976). *The psychology of mathematical ability in school children.* (J. Teller, trans.). Chicago: University of Chicago Press.

Kulik, J. A., & Kulik, C.-L. C. (1984). Effects of accelerated instruction on students. *Review of Educational Research, 54*(3), 409-425.

Kulik, J. A., & Kulik, C-L C. (1987). Effects of ability grouping on student achievement. *Equity & Excellence. 23*(1-2), 22-30.

Lenchner, G. (1983). *Creative problem solving in school mathematics.* Boston: Houghton-Mifflin.

Lupkowski, A. E., & Assouline, S. G. (1992, February). *Identifying mathematically talented elementary students: Using the lower level of the SSAT as an out-of-level test.* Paper presented at the Esther Katz Rosen Symposium on the Psychological Development of Gifted Children, University of Kansas.

References

Lupkowski, A. E., Assouline, S. G., & Stanley, J. C. (1990). Applying a mentor model for young mathematically talented students. *Gifted Child Today, 13*(2), 15-19.

Lupkowski, A. E., Assouline, S. G., & Vestal, J. (1992). Mentors in math. *Gifted Child Today, 15*(3), 26-31.

Marland, S. P., Jr. (1972). *Education of the gifted and talented: Report to the Congress of the United States by the U.S. Commissioner of Education.* Washington, D. C.: U. S. Government Printing Office.

Milgram, R. M. (1991). *Counseling gifted and talented children: A guide for teachers, counselors, and parents.* Norwood, NJ: Ablex.

Mills, C. J., & Wood, S. (1988). *Fast-paced, individualized arithmetic/pre-algebra course, version B: Volume 1, Curriculum Guide.* Baltimore, MD: Center for the Advancement of Academically Talented Youth, Johns Hopkins University.

Moore, N. D., & Wood, S. S. (1988). Mathematics with a gifted difference. *Roeper Review, 10*(4), 231-234.

National Council of Teachers of Mathematics. (1980). *An agenda for action: Recommendations for school mathematics of the 1980's.* Reston, VA: National Council of Teachers of Mathematics.

National Council of Teachers of Mathematics. (1989). *Curriculum and evaluation standards for school mathematics.* Reston, VA: The National Council of Teachers of Mathematics, Inc.

Otis-Lennon School Ability Test. (1989). San Antonio: The Psychological Corporation.

Piaget, J., & Inhelder, B. (1969). *The psychology of the child.* New York: Basic Books.

Porter, A. (1989). A curriculum out of balance: The case of elementary school mathematics. *Educational Researcher, 18*(5), 9-15.

References

Pratscher, S. K., Jones, K. L., & Lamb, C. E., (1982). Differentiating Instruction in mathematics for talented and gifted youngsters. *School Science and Mathematics, 82,* 365-372.

Renzulli, J. S., Reis, S. M. (1991). The schoolwide enrichment model. In N. Colangelo and G. A. Davis (Eds.), *Handbook of gifted education* (pp.111-141). Boston: Allyn and Bacon.

Renzulli, J. S., Reis, S. M., & Smith, L. H. (1981). *The revolving door identification model.* Mansfield Center, CT: Creative Learning Press.

Reys, R. E. (1971). Considerations for teachers using manipulative materials. *Arithmetic Teacher, 18,* 551-558.

Robinson N. M., & Janos P. M. (1987). The contribution of intelligence tests to the understanding of special children. In J. D. Day & J. G. Borkowski (Eds.), *Intelligence and Exceptionality: New directions for theory, assessment, and instructional practices* (pp. 21-56). Norwood, N.J.: Ablex Publishing.

Robinson, N. M. & Robinson, H. B. (1982). The optimal match: Devising the best compromise for the highly gifted student. In D. Feldman (Ed.) *Developmental approaches to giftedness and creativity.* San Francisco: Jossey-Bass.

Sattler, J. M. (1982). *Assessment of children's intelligence and special abilities.* Boston: Allyn and Bacon.

Saul, M. E., Kessler, G. W., Krilov, S., & Zimmerman, L. (1986). *The New York City contest problem book.* Palo Alto: Dale Seymour Publications.

Schunk, D. (1987). Peer models and children's behavioral change. *Review of Educational Research, 57*(2), 149-174.

Secondary School Admission Test Interpretive Guide for School Personnel and Educational Consultants. (1990). Princeton, NJ: Secondary School Admission Test Board.

References

Sirr, P. M. (1984). A proposed system for differentiating elementary mathematics for exceptionally able students. *Gifted Child Quarterly, 28,* 40-44.

Stanley, J. C. (1978). SMPY's DT-PI mentor model: Diagnostic Testing followed by Prescriptive Instruction. *Intellectually Talented Youth Bulletin, 4*(10).

Stanley, J. C. (1979a). How to use a fast-pacing math mentor. *Intellectually Talented Youth Bulletin, 5*(6), 1-2.

Stanley, J. C. (1979b). Identifying and nurturing the intellectually gifted. In W. C. George, S. J. Cohn, and J. C. Stanley (Eds.), *Educating the gifted: Acceleration and enrichment.* Baltimore: The Johns Hopkins University Press.

Stanley, J. C. (1988). Some characteristics of SMPY's "700-800 on SAT-M Before Age 13 Group": Youths who reason **extremely** well mathematically. *Gifted Child Quarterly, 32,* 205-209.

Stanley, J. C., & Benbow, C. P. (1986). Youths who reason exceptionally well mathematically. In R. J. Sternberg and J. E. Davidson (Eds.), *Conceptions of Giftedness,* New York: Cambridge University Press.

Stanley, J. C., Lupkowski, A. E., & Assouline, S. G. (1990). Eight considerations for mathematically talented youth. *Gifted Child Today, 13*(2), 2-4.

Stanley, J. C., & Stanley, B. S. K. (1986). High-school biology, chemistry, or physics learned well in three weeks. *Journal of Research in Science Teaching, 23*(3), 237-250.

Sternberg, R. J. (1986). A triarchic theory of intellectual giftedness. In R. J. Sternberg and J. E. Davidson (Eds.) *Conceptions of Giftedness.* Cambridge: Cambridge University Press.

Talent Search expands: CTY now in all Northeastern states. (1991). *Memberanda, 7*(3), 1. Princeton, NJ: Secondary School Admission Test Board.

References

Tannenbaum, A. (1979). Pre-Sputnik to post-Watergate concern about the gifted. In A.H. Passow (Ed.), *The gifted and the talented: Their education and development* (pp. 5-27). Chicago: University of Chicago Press.

VanTassel-Baska, J. (1984). The Talent Search as an identification model. *Gifted Child Quarterly, 28*(4), 172-176.

Wallace, A. (1986). *The prodigy*. New York: E. P. Dutton.

Wheatley, G. H. (1983). A mathematics curriculum for the gifted and talented. *Gifted Child Quarterly, 27*(2), 77-80.

Wheatley, G. H. (1988). Mathematics curriculum for the gifted. (pp. 252-274). In J. VanTassel-Baska (Ed.), *Comprehensive curriculum for gifted learners*. Boston: Allyn & Bacon.

Whitmore, J. R. (1980). *Giftedness, conflict, and underachievement*. Boston: Allyn and Bacon.

Willoughby, S. S. (1990). *Mathematics education for a changing world*. Alexandria, VA: Association for Supervision and Curriculum Development.

Wolfle, J. A. (1987). Enriching the mathematics program for middle school gifted students, *Roeper Review, 9*, 81-85.

References

Index

a

Accelerating 24-31, 95, 101
Acceleration 4, 6, 18, 24-31, 84, 90, 106
    Acceleration Model 55, 56, 78
    Advantages and Benefits of Acceleration 30-31
Achievement Tests 36-38, 44, 46, 86, 88, 94-95, 112, 119-122
    Comprehensive Testing Program-II (CTP-II) 36, 120
    Cooperative Mathematics Tests 120
    Scholastic Aptitude Tests (SAT) 11, 53, 87, 91, 109, 117, 132
    Sequential Tests of Educational Progress (STEP) 36, 44, 60, 86, 102, 119-120
Advanced Placement Program 133
Algeblaster 94
Algebra 11, 30, 34, 39, 60, 70-71, 85, 87, 90, 94, 129
    Algebra I 45, 50, 56, 90, 94, 99, 109, 132
    Algebra II 50, 91, 109, 132
    Pre-algebra 45, 69, 74, 90, 99, 102, 109, 132
Application of the DT→PI Model 87-92
Aptitude Tests 35-36, 44, 46, 53, 86, 88, 92, 112, 116-119
    Bennett Test of Mechanical Comprehension 118
    Differential Aptitude Test (DAT) 118
    Raven's Progressive Matrices (RPM) 13, 105, 119
    School and College Abilities Test (SCAT) 35, 86, 117
    Secondary School Admissions Test (SSAT) 35, 36, 45, 103, 116-117
Arithmetic and Algebraic Concepts 58, 61, 64, 75
Arizona State 26, 69, 133

Index

b

Basic Concepts 36, 38, 41, 44, 45, 60, 86, 102, 121
BASIC 65, 85, 109
Bennett Test of Mechanical Comprehension 118
Beyond-Level Testing 33, 34, 35, 36, 116-119, 122
Breadth/Depth Model 55, 56, 78

c

Calculators 6, 55, 59, 65, 75
Carnegie Mellon 26
Ceiling Effect 33, 34
Center for the Advancement of Academically Talented Youth (CTY)
    11, 25, 26, 42, 65, 69
Center for Academic Precocity 69
Cognitive Abilities Test 112
Cognitive Developmental Approach 2, 12-13
Comprehensive Testing Program (CTP-II) 36, 120
Computation 38, 40, 41, 44, 45, 60, 86, 102
Computers 55, 59, 64, 65, 66, 72, 82, 93, 136
    Computer Games 24, 71-74
    Computer Programs 52, 71-74
    Computer Software 72-74
    Computer Programming 58, 61, 64-65
        LOGO 65, 74
        BASIC 65, 85, 109
Contests 24, 47, 52, 77-78, 87, 106, 131
    Contest problems 71, 139-156
Cooperative Mathematics 44
Cooperative Mathematics Tests 120
Cooperative Learning 6, 20, 48-49
CTY 11, 25, 26, 42, 65, 69
Cumberland Accelerated Mathematics Project 6
Curriculum 6-7, 55-79
    Curriculum Compacting 22
    Curriculum-Based Assessment 120-121
    Differentiated Curriculum 18-19, 21, 23, 55, 103
    Elementary Curriculum 18
    Mathematics Curriculum 21, 23, 39, 43, 55-80, 85, 102, 131
    Spiral Curriculum 3, 68

Index

d

DAT 118
Diagnostic Testing → Prescriptive Instruction 6, 27, 33-54, 56, 63, 78, 83-87, 87-92, 97, 102, 132
Diagnostic Testing 6, 33-54, 88
Diagnostic/Prescriptive Approach 33, 56, 63
Diagnostic Assessment 101
Differential Aptitude Test (DAT) 118
Differentiated Programming 83
Differentiated Curriculum 18, 21, 23, 55, 103
DT→PI 6, 27, 33-54, 56, 63, 78, 83-87, 87-92, 97, 102, 132
Duke University's Talent Identification Program 11, 133

e

Educational Reform 17
Elementary Mathematics 39, 40, 57-58, 129
Elementary Curriculum 18
Elements of Mathematics (EM) 4
EM 4
Enrichment 3, 5-6, 18, 21-24, 36, 93, 94, 97, 106, 126
    Enrichment Approach 101
    Enrichment Program 5, 18, 22, 24, 94, 98
    Enrichment Model 55, 78
    Enrichment Books 70-71
Estimation 58, 59, 61, 65

f

Facts and Computation 58, 61, 63-64
Fast-Paced Classes 6, 24, 25-26, 42
Flexible School Program 110
Flexible Pacing 6

Index

g

Gardner 2, 14
Geometry 4, 50, 91, 99, 120, 129, 132
Geometry and Measurement 58, 61, 62-63
Grade-Skipping 4, 5, 18, 24, 25, 26, 28, 43, 52, 82
Grouping 17-18, 56, 82

h

Hollingworth, Leta 9-11, 34
Home-schooling 24, 103, 106
Homework 21, 40, 42, 47, 48, 86, 88, 89, 94

i

IEP 98
IMAES 43, 44, 46, 51
Individualized Educational Plan (I.E.P.) 98
Individualized Mathematics Instruction 98
Individually-Paced Programs 6, 24, 27
Information Processing 2, 12, 13-14
    Information-Processing Approach 13-14
Intelligence Tests 112-116
    Kaufman-Assessment Battery for Children (K-ABC) 114
    McCarthy Scales 115
    Stanford-Binet 112-113
    Wechsler Scales 113-114
Intelligence Test Scores 115-116
International Mathematical Olympiad 78, 131
Investigation of Mathematically Advanced Elementary Students 43, 44, 46, 51
Iowa Acceleration Scale 27
Iowa Tests of Basic Skills (ITBS) 34, 35, 44, 60, 92, 101, 117
IQ 2, 12, 29, 92, 104
ITBS 34, 35, 44, 60, 92, 101, 117

j

JCSMP 43, 44, 46, 50, 51, 69-70
Julian C. Stanley Mentor Program (JCSMP) 43, 44, 46, 50, 51, 69-70

k

K-ABC 114
Kaufman Assessment Battery for Children (K-ABC) 114
Key Math Test 93

l

LOGO 65, 74

m

Manipulatives 62, 64, 67, 75-77
Marland Report 2
Mathematics, Elementary 39, 40, 57-58, 129
MathCounts 78, 87, 93, 131
Mathematical 59, 67, 75, 78, 92
Mathematical Maturity 129, 131
Mathematics Enrichment 23-24, 52, 98
Mathematics Curriculum 21, 23, 39, 43, 55-80, 85, 102, 131
Mathematics Games 77
Mathematical Olympiad for Elementary Schools (MOES) 24, 47, 71, 84-85, 87, 93, 106, 131
McCarthy Scales 115
Mentor 24, 27, 33, 38-41, 41-42, 44-48, 51, 53, 86, 87-92, 96-98,
    Mentor-To-Student 46
    Mentor Program 43-44, 46, 88, 92
    Mentor-Paced 42, 46, 51, 52
        Mentor-Paced Programs 33, 43-51, 89-92, 109
        Mentor-Paced Instruction 11, 44
MOES 24, 47, 71, 84-85, 87, 93, 106, 131
Multiple Intelligences 2, 12, 14

Index

n

National Council of Teachers of Mathematics (NCTM) 5, 23, 57, 58-60, 65, 67
NCTM 5, 23, 57, 58-60, 67
Northwestern's Midwest Talent Search 11
Numeration 58, 61, 65-66

O

Otis-Lennon School Ability Test 112
Out-Of-Level Aptitude Test 35, 44, 92, 103, 109, 117

p

Peer Tutoring 21
Percentile ranking 115, 116, 121
Periodicals 66, 77
Piaget 2, 12
    Piagetian stage theory 2, 12, 13
Post-Tests 33, 40, 46, 53
Post-Testing 40-41
Pre-Algebra 45, 69, 74, 90, 99, 102, 109, 132
Precocious 13, 29
Prescriptive Instruction 33-54
Probability and Statistics 58, 61, 66, 71
Problem-solving heuristics 61
Problem solving 5, 6, 18, 22-23, 36, 39, 58, 59, 60, 61-62, 64, 75, 93, 129
Programs 11, 18-27, 43-52, 89-92, 92-93, 102
Psychoeducational Report 122-127
Psychometric Approach 2, 12, 14
Publicity 29, 30

r

Raven's Progressive Matrices (RPM) 13, 105, 119
Recreational Mathematics Books 24, 70-71

RPM 13, 105, 119,

S

Sacramento State 26
SAT 11, 53, 87, 91, 109, 117, 132
SCAT 35, 86, 117
Scholastic Aptitude Test (SAT) 11, 53, 87, 91, 109, 117, 132
School and College Abilities Test (SCAT) 35, 86, 117
Secondary School Admission Test (SSAT) 45, 103, 116-117
    Secondary School Admission Test-Lower Level 35, 44, 101
Sequential Tests of Educational Progress (STEP) 36, 44, 60, 86, 102, 119-120
Set Theory 129, 134
Sets 39, 70, 76
Sidis, William James 29
Slosson Intelligence Test 93
SMPY 11, 27, 33, 34, 35, 41, 55, 69, 95
Social Readiness 28
Social Development 27-29
Spatial Visualization 58, 61, 66
Speyer School 10
Spiral Approach 17, 69
Spiral Curriculum 3, 68
SSAT 35, 44, 45, 101, 103, 116-117
Standard Deviation 12
Standardized Mathematics Tests 36, 94, 96
Stanford-Binet 9, 10, 112-113
Stanley, Julian C. 11, 33, 37, 43, 132
Status Quo 19-20, 98
STEP 36, 44, 60, 86, 102, 119-120
Sternberg, Robert 2, 13-14
    Triarchic theory 13-14
Study of Mathematically Precocious Youth (SMPY) 11, 27, 33, 34, 35, 41, 55, 69, 95

t

Talent Identification Program (TIP) 11, 133
Talent Searches 11, 87, 92, 106, 132
Talented Youth Mathematics Program 11
Tangrams 66, 76
Terman 2, 12, 113

Index

Terman Classes 10
Tessellations 66
Textbooks 39, 45, 62, 64, 65, 68-70
The Science Toolkit 94
The University of Iowa 26, 133
TIP 11
Topics Model 55, 56, 78
Triarchic theory 13-14

u

Unified approach 3-4
University of Minnesota's Talented Youth Mathematics Program 11
University of North Texas 26, 70, 133

w

Wechsler Scales 113-114, 115
    WAIS, WAIS-R, WISC, WISC-III, WPPSI, WPPSI-R 114
    Wechsler Intelligence Scale for Children WISC 114
    Wechsler Preschool and Primary Scale of Intelligence (WPPSI) 114
    Wechsler Intelligence Scale for Children-Revised (WISC-R) 85, 92, 114
    Wechsler Adult Intelligence Scale (WAIS) 114
    WISC-R 85, 92, 114, 116

y

Young Students Program 65

About the Authors...

Ann E. Lupkowski earned a Ph.D. in Educational Psychology at Texas A&M University and then completed a three-year postdoctoral fellowship at the Study of Mathematically Precocious Youth at Johns Hopkins University in 1989. She was an assistant professor and Director of the Study of Mathematically Precocious Youth at the University of North Texas from 1989 to 1991. There she founded the Julian C. Stanley Mentor Program, an individualized, mentor-paced program for mathematically talented elementary students. She is currently Director of the Investigation of Talented Elementary Students at Carnegie Mellon University, where she continues to conduct research and develop programs on behalf of mathematically talented youth. She can be reached at Investigation of Talented Elementary Students, 201 Smith Hall, Pittsburgh, PA 15213.

Susan G. Assouline is Associate Director of The Connie Belin National Center for Gifted Education at The University of Iowa. She is also the clinical supervisor at the Belin Center. Dr. Assouline earned her Ed.S. in school psychology and her Ph.D. in Educational Psychology from The University of Iowa. Upon completion of her doctoral degree, she spent two years at Johns Hopkins University where she had a postdoctoral fellowship with the Study of Mathematically Precocious Youth. She has maintained her interest in working with academically talented youth and co-directs a variety of precollege programs and talent searches for elementary, junior high, and high school students. She can be reached at The Connie Belin National Center for Gifted Education, The University of Iowa, 210 Lindquist Center, Iowa City, IA 52242.